Randomized Response Techniques

Arijit Chaudhuri · Sanghamitra Pal · Dipika Patra

Randomized Response Techniques

Certain Thought-Provoking Aspects

 Springer

Arijit Chaudhuri
Indian Statistical Institute
Kolkata, West Bengal, India

Sanghamitra Pal
West Bengal State University
Kolkata, West Bengal, India

Dipika Patra
Seth Anandram Jaipuria College
Kolkata, West Bengal, India

ISBN 978-981-99-9668-1   ISBN 978-981-99-9669-8   (eBook)
https://doi.org/10.1007/978-981-99-9669-8

Mathematics Subject Classification: 62D05, 62P12, 62P20

This Springer imprint is published by the registered company Springer Nature Singapore Pte Ltd.
The registered company address is: 152 Beach Road, #21-01/04 Gateway East, Singapore 189721,
Singapore

Paper in this product is recyclable.

To Bulu
Arijit Chaudhuri

To my Late Parents
Sanghamitra Pal

To my Parents and Brother
Dipika Patra

Preface

Through authoring the 2011 monograph on randomized response (RR) technique (RRT) and co-authoring one in 1988 and another in 2013 with Rahul Mukerjee and Tasos Christofides and finally co-editing the *Handbook of Statistics* 34 (2016) with Profs. Calyampudi R. Rao and Tasos Christofides and especially while executing the latest task, Arijit Chaudhuri gratefully admits that he became acquainted with a very gratifying community of researchers in this area of research and application. As he thus became familiar with many new ideas and concepts, a few more cropped up in the minds of all three authors who work together in Kolkata, India. Hence, the three co-authors of this monograph decided to write it.

Since Warner's (1965) debut, numerous methods have appeared in the literature on RR techniques. In comparing among their respective device, a standard practice is to consider assigning values to various parameters in respective RR procedures and evaluate measures of errors and also evaluate the measures of protected privacy. We examined how the RR survey data gathered by a procedure of RR data collection based on a specific sampling design promises to fare vis-à-vis another one that might have been employed but not actually implemented. In case of direct survey data (DR survey data), a theory is available in the literature but nothing of the sort has yet been tried in RR situations. In Chaudhuri (2011), it was emphasized that almost every available RRT based on simple random sampling taken with replacement (SRSWR) in practice may easily be extended to alternative ways of sampling by general alternative procedures. So, in this monograph, we present a requisite theory with RRTs combined with alternative sample-selection procedures. We present elaborate numerical illustrations through simulations utilizing live data taken from existing literature. Chapter 3 gives quite a comprehensive detail.

Another exercise missing in the RRT literature is "how to set a reasonable size of a sample in RR surveys". In DR surveys, since Chaudhuri's (2010) publication, it is well known how the celebrated Chebyshev's inequality may be profitably employed to set reasonably the size of a sample planned to be chosen by SRSWR and SRSWOR methods. Through subsequent work by him and his colleagues, it is further known how to extend these ideas to cover the situations when data are to be gathered by other sampling procedures and RR. In Chap. 4, the details are given.

In Chaudhuri and Mukerjee (1988), Chaudhuri (2011), Chaudhuri and Christofides (2013), some ideas even though prevalent were not discussed at all or to an adequate length. We have endeavoured to cover some of them with more diligence. One such idea is the study of "sensitivity level" in optional randomized response (ORR) techniques (ORRT). A detailed study on this has been presented in this monograph. Admittedly, it has not been possible for us to voraciously peruse every important publication on RRT and some of its variants but a major portion of what is available, we have endeavoured to cover to the best of our ability in this monograph.

Especially in Chaudhuri (2011), we clearly tried to announce that RRTs need not be confined to SRSWR alone but may be combined with general sampling procedures but even in the latest publications today, we see RRTs are being confined to the SRSWR alone. It may be surmised that a crucial reason why RRTs are not yet treated as a part of survey sampling as a subject owing to this over emphasis of RRTs being tied to SRSWR.

In our earlier publications, we did not emphasize Franklin's (1989) RRT, which when explained by Marcheselli and Barabesi (2006) has a simple and interesting application and even shows it to be general enough with many popular RRTs as simple special cases of this parent RRT. In this monograph, we tried to mitigate this lapse.

Another important original approach is by Sarjinder Singh and Sedory (2013) which involves Bernoulli trials and hypergeometric approach with trials without replacement and their extension to inverse Bernoulli and inverse hypergeometric trials. An important off-shoot of this Singh-Sedory approach is the study of revised URL model in RRT. This has been amply covered in this volume starting with Chaudhuri and Shaw's (2017) paper.

Kolkata, India   Arijit Chaudhuri
                                                                                    Sanghamitra Pal
                                                                                        Dipika Patra

References

Chaudhuri, A. (2010). *Essentials of survey sampling.* Delhi, India: Prentice Hall of India.

Chaudhuri, A. (2011). *Randomized response and indirect questioning techniques in surveys.* Boca Raton: CRC Press.

Chaudhuri, A., & Christofides, T. C. (2013). *Indirect questioning in sample surveys.* Berlin, Germany: Springer Verlag.

Chaudhuri, A., & Mukerjee, R. (1988). *Randomized responses: theory and techniques.* New York: Marcel Dekker.

Chaudhuri, A. & Shaw, P. (2017). Empirical bayes estimation using quantitative randomized response data. *Statistics and Applications, New Series, 15*(1,2), 1–6.

Chaudhuri, A., Christofides, T. C., & Rao, C. R. (2016). *Handbook of statistics, data gathering, analysis and protection of privacy through randomized response techniques: Qualitative and quantitative human traits.* (Vol. 34). NL: Elsevier.

Franklin, L. A. (1989). A comparision of estimators for randomized response sampling with continuous distributions from dichotomous population. *Communications in Statistics—Theory and Methods, 18*(2), 489–505.

Marcheselli, M., & Barabesi, L. (2006). A generalization of Huang's randomized response procedure for the estimation of population proportion and sensitivity level. *Metron, LXIV*(2), 145–159.

Singh, S., & Sedory, S. A. (2013). A new randomized response device for sensitive characteristics: an application of the negative hypergeometric distribuion. *Metron, 71*(1), 3–8.

Warner, S. L. (1965). Randomized response: a survey technique for eliminating evasive answer bias. *Journal of American Statistical Association, 60*, 63–69.

Acknowledgements

Arijit Chaudhuri gratefully acknowledges help and encouragement received from the Director, Indian Statistical Institute (ISI), and his colleagues in Applied Statistics Unit of ISI. Sanghamitra Pal gratefully acknowledges support and encouragement profusely received from her family. Dipika Patra conveys her gratitude to her parents for the immense help. Also our thanks goes to the reviewers for their suggestions that helped to improve the content.

Contents

About the Authors

Arijit Chaudhuri is Honorary Visiting Professor at the Applied Statistics Unit at the Indian Statistical Institute (ISI), Kolkata, India, since 1st September 2005. He holds a Ph.D. in Statistics in the area of sample surveys from the University of Calcutta, Kolkata, India, from where he also graduated. He worked as a postdoctoral researcher for two years at the University of Sydney (1973–1975), Australia. He retired as a professor from the ISI, Kolkata, India, on 31st August 2002, where he then continued to work as the CSIR Emeritus Scientist for three years up to 31st August 2005. His areas of research include mean square error estimation in multi-stage sampling, analytical study of complex surveys, randomized response surveys and small area estimation. In 2000, he was elected as the President of the Section of Statistics for the Indian Science Congress and worked for the Government of West Bengal for 12 years as the Committee Chairman for the improvement of crop statistics. He has also worked with the Government of India to apply sophisticated methods of small area estimation in improving state and union territory level estimates of various parameters of national importance. He has worked on various global assignments upon invitation, including visiting professorships at several universities and statistical offices in the USA, Canada, England, Germany, the Netherlands, Sweden, Israel, Cyprus, Turkey, Cuba and South Africa, from 1979 to 2009. He has successfully mentored ten Ph.D. students and published more than 150 research papers in peer-reviewed journals, a few of them jointly with his students and colleagues. He is the author of 11 books on survey sampling.

Sanghamitra Pal is Assistant Professor at the Department of Statistics, West Bengal State University (WBSU), India, since 2009. She has completed her Ph.D. in Statistics from the Indian Statistical Institute (ISI), Kolkata, in 2004. Earlier, she served as a research scientist from 2001 to 2009 at River Research Institute (RRI), the Government of West Bengal, India. She also worked as Research Associate and a Visiting Scientist at the Applied Statistics Unit, ISI, Kolkata, from January 2005 to July 2006. She is guiding two Ph.D. students in the area of sample surveys at WBSU and has published research articles in several reputed journals. She organized an invited session on adaptive cluster sampling at the International Statistical

Institute Conference in Durban, South Africa (2009); a sample survey session at the Indian Science Congress (2012); and presented papers at international conferences in South Africa, New Zealand, Germany, France, Thailand, Singapore, as well as India. She was involved in the professional attachment training (PAT) program of sample survey with ICAR-NAARM, Hyderabad, India. Besides teaching and research, she is working as a nodal officer for the All India Survey of Higher Education cell of WBSU.

Dipika Patra is Faculty at Seth Anandram Jaipuria College, Kolkata, West Bengal, India, since August 2014. Moreover, she is involved in postgraduate teaching at the Department of Statistics, West Bengal State University (WBSU), Barasat, India. She earned her Ph.D. in Statistics from WBSU, in 2022. She has more than ten publications in peer-reviewed journals: *Metron* and *Communications in Statistics: Simulation and Computation and Statistics in Transition*. She ranked third in the Bose–Nandi Award from Calcutta Statistical Association, in 2019.

Chapter 1
Genesis, Background and the Need for Randomized Response Techniques (RRT)

1.1 Introduction

Warner (1965) gave us a pioneering ingenious device to easily procure reliable and trustworthy personal facts of sensitive and stigmatizing nature which people are expected to keep secret and hide from others. His celebrated method is known as randomized response (RR) technique (RRT). Habits of drunken driving, under-reporting in Income Tax Returns, hiding history of AIDS, venereal diseases, unfair claims for social reliefs and grants are some of such issues on which it is hard to gather honest responses in usual direct social surveys.

As a possible way out, Warner's (1965) advice was to send an interviewer to a sample-selected person offering him/her a box packed with identical-shaped cards with a proportion p $(0 < p < 1, p \neq \frac{1}{2})$ marked A which represents a sensitive feature as illustrated above and the rest marked A^c which denotes a complementary characteristic. The person responds by randomly picking one card from the box and announcing whether his/her feature A or A^c matches the card type A or A^c picked. Obviously, he/she is not to show the interviewer the type of the card actually drawn.

When this exercise is completed covering quite a few interviewees, it is quite possible to statistically estimate the unknown proportion $\theta (0 \leq \theta \leq 1)$ of the people in the population bearing the feature A or the total of people of our concern. The theme of this book is how to execute this essential investigation appropriately.

An astounding volume of literature has thus far developed since this seemingly simple origin. At least more than a hundred enthusiastic social scientists and statisticians across the world have taken interest and developing this innovative subject since this inception in 1965 through these 55 years with little breaks across time. Numerous papers covering wide ramifications of the subject in various directions have been published in several peer-reviewed journals to date.

The primeval book-length treatise on RRT was penned by Fox and Tracy (1986). Chaudhuri and Mukerjee (1988), Chaudhuri (2011), Chaudhuri and Christofides

© The Author(s), under exclusive license to Springer Nature Singapore Pte Ltd. 2024
A. Chaudhuri et al., *Randomized Response Techniques*,
https://doi.org/10.1007/978-981-99-9669-8_1

(2013) and Fox (2016) followed suit. Handbook of Statistics 34 (2016) edited by Chaudhuri, Christofides and Rao enriched the growing subject further.

We shall conclude our discussions by reviewing the two initial approaches in RRTs.

1.2 Warner's (1965) Device

Let $U = (1, 2, \ldots, i, \ldots, N)$ denote a finite population of N persons and $\underline{Y} = (y_1, y_2, \ldots, y_N)$ an N-vector denoting the values y_i of a real variable y defined on U for its N units ($i = 1, \ldots, N$). For the Warner's (1965) device, each y_i is valued either 1 or 0 so that $Y = \sum_{i=1}^{N} y_i$ is the total number of units in U valued unity or the number of people in the population bearing a sensitive characteristic denoted by A. The population mean $\overline{Y} = \frac{Y}{N}$ in this situation denotes the proportion θ in the population bearing A. The problem addressed by Warner (1965) was to estimate this proportion θ of a qualitative characteristic alone. Warner was a social scientist and not a statistician, and his investigation was restricted to Simple Random Sampling (SRS) With Replacement (WR), i.e. SRSWR of units from U. But there is no need for such a restriction, and his RRT is applicable to general selection procedures as we shall show.

Warner's (1965) device prescribes an interviewer to elicit from a sampled person labelled i an RR as

$$I_i = \begin{cases} 1 & \text{if } i \text{ finds a 'match' in his/her feature } A \text{ or } A^c \text{ with the card type } A \text{ or } A^c \\ 0 & \text{if there is 'no match'.} \end{cases}$$

Denoting by E_R, V_R the operators for expectation, variance with respect generically to any RR device, it follows that

$$E_R(I_i) = py_i + (1 - p)(1 - y_i) = (1 - p) + (2p - 1)y_i$$
$$V_R(I_i) = p(1 - p), \text{ noting } I_i = 1.0; \ y_i = 1.0.$$

Taking $r_i = \frac{I_i - (1-p)}{2p-1}$, it follows that $E_R(r_i) = y_i$ and $V_R(r_i) = \frac{p(1-p)}{(2p-1)^2} = V_i$, say $\forall i \in U$.

Let s be a sample of units selected from U with a probability $P(s)$ according to a design P. Let, further, $\pi_i = \sum_{s \ni i} P(s)$ and $\pi_{ij} = \sum_{s \ni i, j} P(s)$ denote, respectively, the inclusion probability of i in a sample chosen according to design P and the joint inclusion probability of the units i and j of U according to the design P. We shall assume throughout that $\pi_i > 0 \, \forall i \in U$ and $\pi_{ij} > 0 \, \forall i, j \in U$.

Then, we may take the Horvitz-Thompson estimator

$$t_{\text{HT}} = \sum_{i \in s} \frac{y_i}{\pi_i}$$

as an unbiased estimator for Y having variance

$$V_p(t_{\mathrm{HT}}) = \sum_{i=1}^{N} \sum_{\substack{j=1 \\ i<j}}^{N} (\pi_i \pi_j - \pi_{ij})(\frac{y_i}{\pi_i} - \frac{y_j}{\pi_j})^2$$

Let us denote by E_P, V_P the expectation, variance operators with respect to the sampling design P. Here, we assume every sample has all its units distinct, else the variance above would have one more term $\sum_{i=1}^{N} \frac{y_i^2}{\pi_i} \beta_i$ with $\beta_i = 1 + \frac{1}{\pi_i} \sum_{j\neq i=1}^{N} \pi_{ij} - \sum_{i=1}^{N} \pi_i$. We keep it simple here. By $E = E_P E_R = E_R E_P$, we shall denote the overall expectation operator according to a design and an RR procedure. These expectations are commutative. Also, the overall variance operator according to a design P and generically an RR device as $V = E_P V_R + V_P E_R = E_R V_P + V_R E_P$. Also, an unbiased estimator for $V_P(t_{\mathrm{HT}})$ is

$$v_P(t_{\mathrm{HT}}) = \sum_{i<j} \sum_{\in s} \left(\frac{\pi_i \pi_j - \pi_{ij}}{\pi_{ij}} \right)(\frac{y_i}{\pi_i} - \frac{y_j}{\pi_j})^2.$$

Let us turn to the important aspects of RRT. Note that

$$e_{\mathrm{HT}} = \sum_{i \in s} \frac{r_i}{\pi_i}$$

has $E_R(e_{\mathrm{HT}}) = t_{\mathrm{HT}}$, $E_p(e_{\mathrm{HT}}) = \sum_{i=1}^{N} r_i = R$, say, and $E(e_{\mathrm{HT}}) = E_p E_R(e_{\mathrm{HT}}) = E_p(t_{\mathrm{HT}}) = Y$ and $E(e_{\mathrm{HT}}) = E_R E_p(e_{\mathrm{HT}}) = E_R(R) = \sum_{i=1}^{N} y_i = Y$, i.e. e_{HT} is an unbiased estimator for Y. Also,

$$V(e_{\mathrm{HT}}) = E_p V_R(e_{\mathrm{HT}}) + V_p E_R(e_{\mathrm{HT}})$$

$$= E_p \left(\sum_{i \in s} \frac{V_i}{\pi_i^2} \right) + V_p(t_{\mathrm{HT}})$$

$$= \sum_{i=1}^{N} \frac{V_i}{\pi_i} + \sum_{i=1}^{N} \sum_{\substack{j=1 \\ i<j}}^{N} (\pi_i \pi_j - \pi_{ij}) \left(\frac{y_i}{\pi_i} - \frac{y_j}{\pi_j} \right)^2$$

$$= \frac{p(1-p)}{(2p-1)^2} \sum_{i=1}^{N} \frac{1}{\pi_i} + \sum_{i=1}^{N} \sum_{\substack{j=1 \\ i<j}}^{N} (\pi_i \pi_j - \pi_{ij}) \left(\frac{y_i}{\pi_i} - \frac{y_j}{\pi_j} \right)^2 \quad (1.1)$$

To find an unbiased estimator for $V(e_{\mathrm{HT}})$, let us consider

$$v(e_{\mathrm{HT}}) = \frac{p(1-p)}{(2p-1)^2} \sum_{i \in s} \frac{1}{\pi_i} + \sum_{i<j} \sum_{\in s} \frac{(\pi_i \pi_j - \pi_{ij})}{\pi_{ij}} \left(\frac{r_i}{\pi_i} - \frac{r_j}{\pi_j} \right)^2. \qquad (1.2)$$

Then,

$$E(v(e_{\mathrm{HT}})) = E_p E_R(v(e_{\mathrm{HT}}))$$

$$= E_P E_R \left(\frac{p(1-p)}{(2p-1)^2} \sum_{i \in s} \frac{1}{\pi_i} + \sum_{i<j} \sum_{\in s} \frac{(\pi_i \pi_j - \pi_{ij})}{\pi_{ij}} \left(\frac{r_i}{\pi_i} - \frac{r_j}{\pi_j} \right)^2 \right)$$

$$= \frac{Np(1-p)}{(2p-1)^2} + E_P \left(\sum_{i<j} \sum_{\in s} \frac{(\pi_i \pi_j - \pi_{ij})}{\pi_{ij}} \left(\left(\frac{y_i}{\pi_i} - \frac{y_j}{\pi_j} \right)^2 + \frac{V_i}{\pi_i^2} + \frac{V_j}{\pi_j^2} \right) \right)$$

$$= \frac{Np(1-p)}{(2p-1)^2} + E_P \left(\sum_{i<j} \sum_{\in s} \frac{(\pi_i \pi_j - \pi_{ij})}{\pi_{ij}} \left(\frac{y_i}{\pi_i} - \frac{y_j}{\pi_j} \right)^2 \right)$$

$$+ E_P \left(\sum_{i<j} \sum_{\in s} \frac{(\pi_i \pi_j - \pi_{ij})}{\pi_{ij}} \left(\frac{V_i}{\pi_i^2} + \frac{V_j}{\pi_j^2} \right) \right)$$

$$= \frac{Np(1-p)}{(2p-1)^2} + \sum_{i=1}^{N} \sum_{\substack{j=1 \\ i<j}}^{N} (\pi_i \pi_j - \pi_{ij})(\frac{y_i}{\pi_i} - \frac{y_j}{\pi_j})^2$$

$$+ \sum_{\substack{i \neq j}}^{N} \sum_{=1}^{N} (\pi_i \pi_j - \pi_{ij}) \frac{V_i}{\pi_i^2}$$

$$= \sum_{i=1}^{N} \sum_{\substack{j=1 \\ i<j}}^{N} (\pi_i \pi_j - \pi_{ij}) \left(\frac{y_i}{\pi_i} - \frac{y_j}{\pi_j} \right)^2 + \sum_{i=1}^{N} \frac{V_i}{\pi_i}$$

$$= V(e_{\mathrm{HT}})$$

as

$$\sum_{\substack{i \neq j}}^{N} \sum_{=1}^{N} (\pi_i \pi_j - \pi_{ij}) \frac{V_i}{\pi_i^2} = \sum_{i=1}^{N} \frac{V_i}{\pi_i}(n - \pi_i) - \sum_{i=1}^{N} \frac{V_i}{\pi_i^2}(n-1)\pi_i$$

$$= n \sum_{i=1}^{N} \frac{V_i}{\pi_i} - \sum_{i=1}^{N} V_i - n \sum_{i=1}^{N} \frac{V_i}{\pi_i} + \sum_{i=1}^{N} \frac{V_i}{\pi_i}$$

$$= \sum_{i=1}^{N} \frac{V_i}{\pi_i} - \sum_{i=1}^{N} V_i$$

if each sample s has n units.

Thus, the above $v(e_{HT})$ is an unbiased estimator for $V(e_{HT})$.

Case I:

If an SRSWOR of size n is taken from U, then, from it the

$$e_{HT} \text{ is } \frac{N}{n} \sum_{i \in s} r_i = N\bar{r},$$

writing $\bar{r} = \frac{1}{n} \sum_{i \in s} r_i$, the sample mean of r_i's for i in s because for SRSWOR in n draws $\pi_i = \frac{n}{N} \forall i \in U$.

Also for SRSWOR in n draws, $\pi_{ij} = \frac{n(n-1)}{N(N-1)}$. So, from Eq. 1.1, we have

$$V(N\bar{r}) = \left[\frac{p(1-p)}{(2p-1)^2} \sum_{i=1}^{N} \frac{1}{\pi_i} + \sum_{i=1}^{N} \sum_{\substack{j=1 \\ j<i}}^{N} (\pi_i \pi_j - \pi_{ij}) \left(\frac{y_i}{\pi_i} - \frac{y_j}{\pi_j} \right)^2 \right]$$

$$= \frac{p(1-p)}{(2p-1)^2} \left(\frac{N^2}{n} \right) + \left(N^2 \left(\frac{N-n}{Nn} \right) \right) \frac{1}{N-1} \sum_{i=1}^{N} (y_i - \bar{Y})^2$$

$$= \frac{N^2 p(1-p)}{n(2p-1)^2} + \frac{N(N-n)}{n(N-1)} N\bar{Y}(1-\bar{Y})$$

$$= \frac{N^2 p(1-p)}{n(2p-1)^2} + \frac{N^2(N-n)}{(N-1)} \frac{\theta(1-\theta)}{n}$$

(recalling $\theta = \bar{Y}$)

$$= N^2 \left[\frac{p(1-p)}{n(2p-1)^2} + \frac{(N-n)}{(N-1)} \frac{\theta(1-\theta)}{n} \right].$$

So, an unbiased estimator for the population proportion θ is the sample proportion $\bar{r}$ and its variance is

$$V(\bar{r}) = \frac{p(1-p)}{n(2p-1)^2} + \frac{(N-n)}{(N-1)} \frac{\theta(1-\theta)}{n}.$$

An unbiased estimator for $V(\bar{r})$ is

$$v(\bar{r}) = \frac{p(1-p)}{n(2p-1)^2} + \left(\frac{N-n}{Nn} \right) \frac{1}{n-1} \sum_{i \in s} (r_i - \bar{r})^2.$$

Case II:

If an SRSWR is taken in n draws, then the sample mean $\bar{r} = \frac{1}{n}\sum_{u=1}^{n} r_u$, r_u denoting the value of r_i if i^{th} unit appears on the uth draw ($u = 1, 2, \ldots, n$).
  Then,

$$E(\bar{r}) = E_p\left[\frac{1}{n}\sum_{u=1}^{n} y_u\right] = \frac{1}{N}\sum_{i=1}^{N} y_i = \bar{Y} = \theta.$$

i.e. $\bar{r}$ is an unbiased estimator for θ.
  Now,

$$V(\bar{r}) = E_p V_R(\bar{r}) + V_p E_R(\bar{r})$$
$$= E_p\left[\frac{1}{n^2}\sum_{u=1}^{n}\frac{p(1-p)}{(2p-1)^2}\right] + V_p\left(\frac{1}{n}\sum_{u=1}^{n} y_u\right)$$
$$= \frac{p(1-p)}{n(2p-1)^2} + \frac{1}{Nn}\sum_{i=1}^{N}(y_i - \bar{Y})^2$$
$$= \frac{p(1-p)}{n(2p-1)^2} + \frac{\theta(1-\theta)}{n}$$
$$= \frac{1}{n}\left(\frac{p(1-p)}{(2p-1)^2} + \theta(1-\theta)\right).$$

This formula matches with Warner's (1965) original formula.
  An unbiased estimator of this $V(\bar{r})$ is

$$v(\bar{r}) = \frac{1}{n}\left(\frac{p(1-p)}{(2p-1)^2} + \frac{\hat{\theta}(1-\hat{\theta})}{n-1}\right)$$
$$= \frac{p(1-p)}{n(2p-1)^2} + \frac{\bar{r}(1-\bar{r})}{n-1}$$

which also coincides with Warner's (1965) formula.
  It is well known that for direct response, the estimator $\frac{1}{n}\sum_{i=1}^{n} y_i = \bar{y}$ is not as good as the estimator based on m number of distinct units found in SRSWR sample of size n. This raised the question in RR about a possible comparison between the estimators $\bar{r} = \frac{1}{n}\sum_{i=1}^{n} r_i$ and $\bar{r}\prime = \frac{1}{m}\sum_{i\in s_m} r_i$ where s_m is the set of m distinct units of a sample s of size n chosen by SRSWR. This has been discussed in details in the book by Chaudhuri (2011, pp. 11–35). We should avoid repetition. Also, in RRT, literature use of mean of distinct units in SRSWR sample is not in common practice.

1.3 Unrelated Response Model

Though Warner (1965) did not clearly say it in his treatment if the characteristic A was sensitive and stigmatizing but not so was A^c, rather it was innocuous. But some followers of his pioneering exploits quite critically felt, sometimes A and A^c may both simultaneously be stigmatizing. For example, if A represents supporting a particular political affiliation, A^c as opposing it may as will be sensitive too. So, Simmons, vide Horvitz et al. (1967) and Greenberg et al. (1969) introduced a remarkable alternative to Warner's approach as a new RRT, now famous as an 'unrelated question model' or UQM in brief.

According to this revised RRT, an investigator approaches a sample-selected person i of U with two boxes; both filled with large numbers of cards, in the first marked A and B in proportions $p_1 : (1 - p_1), (0 < p_1 < 1)$ and the other box containing numerous similar cards marked A and B in proportions $p_2 : (1 - p_2), (0 < p_2 < 1$, but $p_1 \neq p_2)$. Here, A is a sensitive feature but B is an innocuous feature, say, preferring B denotes music to painting and B^c denoting its converse, both unrelated to A, A^c which may represent mal-treating one's spouse and its reverse. As with Warner's, y_i equals 1 or 0 to represent A or A^c and x_i equals 1 or 0 to denote bearing the feature B or its converse B^c. Needless to mention, A, A^c, B, B^c all denote qualitative characteristics.

The two boxes being presented to a sampled person labelled i RRs are

$$I_i = 1 \text{ if } i\text{'s feature } A \text{ or } B \text{ matches the card type } A, B \text{ from 1st box}$$
$$= 0 \text{ if it 'does not match'.}$$

and

$$J_i = 1 \text{ if } i\text{'s feature } A \text{ or } B \text{ matches the card type } A, B \text{ taken from 2nd box}$$
$$= 0 \text{ if it 'does not match'.}$$

Thus,

$$E_R(I_i) = p_1 y_i + (1 - p_1)x_i \text{ and } E_R(J_i) = p_2 y_i + (1 - p_2)x_i.$$
$$V_R(I_i) = E_R(I_i)(1 - E_R(I_i)) = p_1(1 - p_1)(y_i - x_i)^2$$
$$V_R(J_i) = E_R(J_i)(1 - E_R(J_i)) = p_2(1 - p_2)(y_i - x_i)^2.$$

Letting

$$r_i = \frac{(1 - p_2)I_i - (1 - p_1)J_i}{p_1 - p_2},$$

one gets $E_R(r_i) = y_i$ and

$$V_R(r_i) = \frac{(1 - p_1)(1 - p_2)(p_1 + p_2 - 2p_1 p_2)}{(p_1 - p_2)^2}(y_i - x_i)^2 = V_i(\text{say}).$$

If a sample s is drawn by a sampling design p such that every sample has the same number n of units, each distinct, such that $\pi_i = \sum_{s \ni i} p(s)$, $\pi_{ij} = \sum_{s \ni i,j} p(s)$ and every π_i and every π_{ij} be positive, then as in the case of Warner's RRT, for this UQM RRT also, we may employ

$$e_{\text{HT}} = \sum_{i \in s} \frac{r_i}{\pi_i}$$

as an unbiased estimator for $Y = \sum_{i=1}^{N} y_i$ with its variance as

$$V(e_{\text{HT}}) = E_p\left(\sum_{i \in s} \frac{V_i}{\pi_i^2}\right) + V_p\left(\sum_{i \in s} \frac{y_i}{\pi_i}\right)$$
$$= \sum_{i=1}^{N} \frac{V_i}{\pi_i} + \sum_{i}^{N}\sum_{<j}^{N}(\pi_i \pi_j - \pi_{ij})\left(\frac{y_i}{\pi_i} - \frac{y_j}{\pi_j}\right)^2.$$

The above noted variance formula given by Yates and Grundy (1953) is applicable here.

$$\text{Since } V_i = V_R(r_i) = E_R\left(r_i^2\right) - (E_R(r_i))^2 = E_R\left(r_i^2\right) - y_i^2$$
$$= E_R\left(r_i^2\right) - y_i = E_R\left(r_i^2\right) - E_R(r_i) = E_R(r_i(r_i - 1)),$$

one may take for V_i an unbiased estimator $v_i = r_i(r_i - 1)$. Then, one may take for $V(e_{\text{HT}})$ an unbiased estimator as

$$v(e_{\text{HT}}) = \sum_{i \in s} \frac{v_i}{\pi_i} + \sum_{i<j}\sum_{\in s}\left(\frac{\pi_i \pi_j - \pi_{ij}}{\pi_{ij}}\right)\left(\frac{r_i}{\pi_i} - \frac{r_j}{\pi_j}\right)^2.$$

For samples s taken by other sampling schemes like SRSWR, SRSWORs also similarly unbiased estimator for Y, $\theta = \frac{Y}{N}$ along with variance formula and unbiased estimator for variances of unbiased estimators may easily be obtained.

References

Chaudhuri, A. (2011). *Randomized response and indirect questioning techniques in surveys.* CRC Press.

Chaudhuri, A., & Christofides, T. C. (2013). *Indirect questioning in sample surveys.* Springer Berlin, Heidelberg.

Chaudhuri, A., & Mukerjee, R. (1988). *Randomized responses: Theory and techniques.* Marcel Dekker.

Chaudhuri, A., Christofides, T. C., & Rao, C. R. (2016). *Handbook of statistics, data gathering, analysis and protection of privacy through randomized response techniques: Qualitative and quantitative human traits* (Vol. 34). Elsevier.

Fox, J. A. (2016). *Randomized response and related methods: Surveying sensitive data.* Sage.

Fox, J. A., & Tracy, P. E. (1986). *Randomized response: A method of sensitive surveys.* Sage.

Greenberg, B. G., Abul-Ela, A. L., Simmons, W. R., & Horvitz, D. G. (1969). The unrelated question randomized response model: Theoretical framework. *Journal of American Statistical Association, 64,* 520–539.

Horvitz, D. G., Shah, B. V., & Simmons, W. R. (1967). The unrelated question randomized response model. In *Proceedings of Social Statistics Section* (pp. 65–72). American Statistical Association.

Yates, F., & Grundy, P. (1953). Selection without replacement from within strata with probability proportional to size. *Journal of the Royal Statistical Society: Series B, 15,* 235–261.

Warner, S. L. (1965). Randomized response: A survey technique for eliminating evasive answer bias. *Journal of American Statistical Association, 60,* 63–69.

Chapter 2
Reviews of Background Material on RRT

2.1 Introduction

Fox and Tracy (1986) had their treatise on RRT written covering merely the early growth of the subject. Sample selection by the researchers till their time was exclusively by the simplest method of Simple Random Sampling With Replacement (SRSWR). Chaudhuri and Mukerjee (1988) pursued essentially the same but only Chap. 7 therein dealt with varying probability sample-selection procedures as well as without replacement. Moreover, this monograph showed how certain classical theoretical developments in the area of direct response (DR) surveys could be extended with requisite modifications to the area of RR surveys. Chaudhuri's (2011) monograph was a real advancement in the realm of RRTs, elaborating on how the classical RR devices could mostly be employed in combination with general sample-selection schemes with equal and varying probabilities with and without replacement. This monograph also dealt with optional RRTs covering qualitative as well as quantitative characteristics. Also, this discussed rational ways to devise RRTs protecting respondents' privacies. Certain alternative devices other than RRTs were also mentioned as capable of protecting privacy and confidentiality in responses. Chaudhuri and Christofides (2013) provided amendments to avoid theoretical preferences in Chaudhuri's (2011) expositions showing more practical aspects of RRTs and citing possibilities of applications. Fox (2016) has written another book on problems of tackling sensitive data collection. This mainly discusses practical procedures of sensitive collection exposing little interest in theoretical aspects of developing refinements in RRTs. So, the onus is on us to convince our readers why they should spare their time and thoughts to care for this volume in addition to the preceding ones.

2.2 Motivation

Let us offer some excuses why we intend to draw your attention to this particular piece. Classically, different devices for gathering RRTs are compared in terms of magnitudes of variances of unbiased estimators for the estimand parameters like totals, means and proportions and the magnitudes of parameters denoting measures of protections of privacy. The smaller the variances the less are the privacy protection measures, are theoretical concepts. They cannot be concretely assessed in terms of values that can be gathered at hand.

In Chap. 3, we introduce a topic as a novelty in the context of RR. Let an RR survey data be obtained by a particular RR device using a specific sampling scheme. Its efficiency can be compared to a rival combination of an RR device and another sampling scheme that might be employed to apply the specific RR device. Several such combinations have been studied as reported in Chap. 3 in this monograph.

Another innovation in the context of RRTs has been presented in Chap. 4. In the context of DR surveys several papers by Chaudhuri and Dutta (2018), Chaudhuri and Sen (2020) and Chaudhuri (2020), procedures applying Chebyshev's inequality in determining sample-sizes in DR surveys and also cursorily in RR surveys have recently appeared. But in Chap. 4 here, this topic has been rather comprehensively dealt with.

After Chaudhuri and Christofides (2013), emerged Handbook of Statistics 34 was published in 2016. It contains articles by numerous authors who published many articles before 2011. Yet the two monographs out in 2011 and 2013 do not mention many of these works. For example, L. Barabesi, A. Quatember, P. F. Perri, M. Rueda among others have contributions worth wide studies. So, in the present monograph we illustratively mention some, which of these works deserve at least cursory coverage and should not be missed any further. So, the three authors of this book have a motivation for incorporating some of these relevant thoughts in this volume.

2.3 Qualitative and Quantitative Characteristics

Warner (1965) started his RR technique considering only a qualitative characteristic, and he considered estimation of a population proportion of people bearing a sensitive characteristic. For an exceedingly long period, investigation continued to concentrate exclusively on a qualitative feature alone. But gradually attention turned to estimating quantitative parameters as well like money lost and won on gambling, cost of treatment of AIDS, amount spent on paying fines and meeting cost of litigation concerning criminal detection, amount under-paid on Income Tax liabilities, etc.

2.4   Parameters to Estimate

Classical interest was principally to estimate proportion of people bearing a sensitive or stigmatizing characteristic. Next interest shifted to estimate loss or gain in the aggregate for gambling over a period, number of days in detention for criminal offence, amounts of bribes paid or earned, cost of fine paid for torturing spouses, etc. Next a new parameter started attracting attention. While executing RR surveys, experience was gathered in course of time that while an investigator treated an item of inquiry as stigmatizing, a section of interviewees felt it rather innocuous like alcoholic habits, uncontrolled driving, not using a helmet while driving a motorbike, etc. In such situations, some of the interviewees may offer to divulge truths about certain items for which an RR device was planned by the investigator. In such a situation, a willing respondent may be permitted to exercise the option to reveal the truth about an item contemplated to be sensitive by the investigator. Then such interviewees may opt to give out the truth directly. A few experts in RRTs then concerned with estimating a new parameter called 'the sensitivity level'. By this is meant the proportion out of those who opt to offer DRs happen to divulge the truth on a characteristic which the investigator treats as stigmatizing. L. Barabesi is a conspicuous expert who has noteworthy contributions to this aspect in RRTs. By its very nature, this is a ratio parameter, and a ratio estimator is naturally proposed to estimate it. This naturally introduces a biased estimator to the RRT literature.

2.5   Devices to Generate RR Data

So far we have talked about only two RR devices given by Warner (1965) and Simmons and his four collaborators as we described in Chap. 1. Many more devices are reported in the literature. A few are described also by Chaudhuri and Mukerjee (1988), Chaudhuri (2011) and Chaudhuri and Christofides (2013). We shall discuss a few here though they involve repetitions because we have to put them to use with specific objectives and motivations. Their respective relevance to be clarified in successive chapters.

Kuk's RR Device

Kuk's (1990) RRT is implemented by an interviewer who approaches a sample-selected person labelled i with two boxes containing quite a large number of similar cards marked red and non-red in proportions $\theta_1 : (1 - \theta_1)$ in the first and $\theta_2 : (1 - \theta_2)$, $(0 < \theta_1 < 1, 0 < \theta_2 < 1, \theta_1 \neq \theta_2)$ in the second box. If i bears A, he/she is to choose by SRSWR in $k(> 1)$ draws a card from the 1st box but from the 2nd box, if he/she instead bears A^c, the complement of A and without divulging the box used to report the number of red cards f_i among the cards in k draws. Then,

$$\text{taking } y_i = 1 \text{ if } i \text{ bears } A$$
$$= 0 \text{ if } i \text{ bears } A^c,$$

$$E_R(f_i) = k[y_i\theta_1 + (1 - y_i)\theta_2] = k[y_i(\theta_1 - \theta_2) + \theta_2] \text{ and}$$
$$V_R(f_i) = k[y_i\theta_1(1 - \theta_1) + (1 - y_i)\theta_2(1 - \theta_2)]$$
$$= k[\theta_2(1 - \theta_2) + y_i\{\theta_1(1 - \theta_1) - \theta_2(1 - \theta_2)\}],$$

since f_i is binomially distributed with parameters k and either θ_1 or θ_2.

Then, $r_i(k) = \frac{\frac{f_i}{k} - \theta_2}{(\theta_1 - \theta_2)}$ has $E_R(r_i(k)) = y_i$ and.

$$V_R(r_i(k)) = \frac{V_R\left(\frac{f_i}{k}\right)}{(\theta_1 - \theta_2)^2} = V_i(k) = b_i(k)y_i + c_i(k) \text{ where } b_i(k) = \frac{1 - \theta_1 - \theta_2}{k^2(\theta_1 - \theta_2)^2} \text{ and}$$
$c_i(k) = \frac{\theta_2(1 - \theta_2)}{k^2(\theta_1 - \theta_2)^2}.$

Estimation of Y unbiasedly is easy employing the Horvitz-Thompson estimator.

$$e_{\text{HT}} = \sum_{i \in s} \frac{r_i(k)}{\pi_i},$$

on choosing a sample according to a design p having $\pi_i > 0 \forall i$ in U. Its variance.

$$V(e_{\text{HT}}) = \sum_{i=1}^{N} \frac{V_i(k)}{\pi_i} + \sum_{i<j}^{N} \sum_{=1}^{N} (\pi_i\pi_j - \pi_{ij})\left(\frac{y_i}{\pi_i} - \frac{y_j}{\pi_j}\right)^2.$$

An unbiased estimator for this $V(e_{HT})$ may be easily checked to be

$$v(e_{\text{HT}}) = \sum_{i \in s} \frac{v_i}{\pi_i} + \sum_{i<j} \sum_{\in s} \left(\frac{\pi_i\pi_j - \pi_{ij}}{\pi_{ij}}\right)\left(\frac{r_i(k)}{\pi_i} - \frac{r_j(k)}{\pi_j}\right)^2$$

provided the design p admits $\pi_{ij} > 0 \forall i \neq j$ in U. Here, $v_i = b_i(k)r_i(k) + c_i(k)$ is an unbiased estimator for $V_i(k)$.

Forced Response Device

Next, we may consider another RR data generating procedure as 'Forced Response Device'. Here an interviewer approaches a probability sampling-based selected person labelled i with a box of a large number of cards marked A, A^c and 'genuine' in respective proportions p_1, p_2 and $1 - p_1 - p_2(0 < p_1 < 1, 0 < p_2 < 1, p_1 \neq p_2$ and $p_1 + p_2 < 1)$.

Then, on request, an RR from i is to be

$$I_i = \begin{cases} 1 & \text{if an 'A' or a 'genuine' marked card is drawn} \\ & \text{and the person's feature is really 'A'} \\ 0 & \text{if an } A^c \text{ card is picked or a 'genuine' marked} \\ & \text{is picked but his/her true feature is } A^c. \end{cases}$$

Then, $E_R(I_i) = \text{Prob}(I_i = 1) = p_1 + y_i(1 - p_1 - p_2)$.

Letting, $r_i = \frac{I_i - p_1}{1 - p_1 - p_2}$, one gets $E_R(r_i) = y_i$ and $V_R(r_i) = \frac{p_1(1-p_1)+y_i(1-p_1-p_2)(p_2-p_1)}{(1-p_1-p_2)^2} = V_i$, say.

Then, $v_i = \frac{p_1(1-p_1)+r_i(1-p_1-p_2)(p_2-p_1)}{(1-p_1-p_2)^2}$ is an unbiased estimator for V_i above.

More RR devices will be described later when relevant theoretical and practical necessities will arise in later chapters. But in the above RR devices, only qualitative features are addressed. But when a quantitative characteristic is to be studied, then we may employ the following device called 'Device I'.

Device I: Quantitative Response

The interviewer is to approach a person labelled i when selected by a probability sampling scheme, with two packs of large numbers of cards, of which the first one is to contain cards bearing numbers $a_1, \ldots, a_j, \ldots, a_T$ and the other pack containing numbers $b_1, \ldots b_l, \ldots, b_K$. On request, the ith person is to draw a card randomly from the 1st and independently from the other pack randomly to pick up one card. Then, the requested RR from i will be

$$I_i = a_j y_i + b_l, \quad j = 1, \ldots, T, \quad l = 1, \ldots, K.$$

Writing $\mu_a = \frac{1}{T}\sum_{j=1}^{T} a_j \neq 0$, $\sigma_a^2 = \frac{1}{T-1}\sum_{j=1}^{T}(a_j - \mu_a)^2$ and $\mu_b = \frac{1}{K}\sum_{l=1}^{K} b_l$, $\sigma_b^2 = \frac{1}{K-1}\sum_{l=1}^{K}(b_l - \mu_b)^2$, one may note.

$E_R(I_i) = y_i \mu_a + \mu_b$,

$$r_i = \frac{I_i - \mu_b}{\mu_a} \text{ has } E_R(r_i) = y_i.$$

$$V_R(r_i) = y_i^2 \frac{\sigma_a^2}{\mu_a^2} + \frac{\sigma_b^2}{\mu_a^2} = V_i, \text{ say.}$$

Then, an unbiased estimator for V_i is $v_i = \left[r_i^2 \frac{\sigma_a^2}{\mu_a^2} + \frac{\sigma_b^2}{\mu_a^2} \right] / (1 + \frac{\sigma_a^2}{\mu_a^2})$.

Device II: Quantitative

Another RR device to produce quantitative data was given by Eriksson (1973). To apply this, an interviewer is to approach a person labelled i selected by a probability sampling scheme with a box of numerous similar cards of which a proportion $C(0 < C < 1)$ of cards is labelled 'true' and the other cards in the pack are labelled real numbers $x_1, x_2, \ldots, x_j, \ldots, x_m$ in respective positive proportions $Q_1, \ldots, Q_j, \ldots, Q_m$ such that $\sum_{j=1}^{m} Q_j = 1 - C$. Then, the RR from i on request to randomly pick just one card from the pack will be

$$I_i = y_i \text{ if a 'true'-marked card is drawn}$$
$$= x_j \text{ if an '}x_j\text{'-marked card is picked up.}$$

Then, $E_R(I_i) = Cy_i + \sum_{j=1}^{m} Q_j x_j$.

and hence $r_i = \frac{I_i - \sum_{j=1}^{m} Q_j x_j}{C}$ has $E_R(r_i) = y_i$ and

$$V_R(r_i) = \frac{1}{C^2} V_R(I_i)$$

$$= \frac{1}{C^2} \left[E_R(I_i^2) - E_R^2(I_i) \right]$$

$$= \frac{1}{C^2} \left[Cy_i^2 + \sum_{j=1}^{m} Q_j x_j^2 - \left(Cy_i + \sum_{j=1}^{m} Q_j x_j \right)^2 \right]$$

$$= \frac{1}{C^2} \left[C(1-C)y_i^2 - 2y_i C \left(\sum_{j=1}^{m} Q_j x_j \right) + \sum_{j=1}^{m} Q_j x_j^2 - \left(\sum_{j=1}^{m} Q_j x_j \right)^2 \right]$$

$$= ay_i^2 + by_i + F = V_i, \text{ say}$$

with a, b, F specified by collecting the coefficients of y_i^2, y_i and the constant terms free of y_i's.

Then, an unbiased estimator for V_i may be worked out on writing $v_i' = ar_i^2 + br_i + F$ such that.

$$E_R(v_i') = a(V_i + y_i^2) + by_i + F = aV_i + [ay_i^2 + by_i + F] = aV_i + V_i \text{ so that}$$

$v_i = \frac{v_i'}{1+a}$ may be taken as an unbiased estimator for $V_i, i \in U$, since $1 + a \neq 0$.

For both these quantitative RR devices for $Y = \sum_{i=1}^{N} y_i$, one may employ Horvitz-Thompson estimator

$$e_{HT} = \sum_{i \in s} \frac{r_i}{\pi_i}$$

for which

$$V(e_{HT}) = \sum_{i=1}^{N} \frac{V_i}{\pi_i} + \sum_{i<j}^{N} \sum_{=1}^{N} (\pi_i \pi_j - \pi_{ij}) \left(\frac{y_i}{\pi_i} - \frac{y_j}{\pi_j} \right)^2.$$

Assuming for simplicity, in every sample, a fixed number n of distinct units and $\pi_i > 0 \, \forall \, i \in U$. Taking in addition $\pi_{ij} > 0 \, \forall \, i \neq j$ in U, an unbiased estimator for $V(e_{HT})$ follows as:

$$v(e_{HT}) = \sum_{i \in s} \frac{v_i}{\pi_i} + \sum_{i<j} \sum_{\in s} \left(\frac{\pi_i \pi_j - \pi_{ij}}{\pi_{ij}} \right) \left(\frac{r_i}{\pi_i} - \frac{r_j}{\pi_j} \right)^2.$$

2.6 Sample-Selection Procedures—Equal and Varying Probabilities

Equal Probability Sampling

Two equal probability sample-selection procedures are well known called Simple Random Sampling With Replacement (SRSWR) and Simple Random Sampling Without Replacement (SRSWOR). In SRSWR, independently from $U = (1, 2, \ldots, i, \ldots, N)$ on each of $n(\geq 1)$ draws, one unit is selected with an equal probability $\frac{1}{N}$. In SRSWOR, it is noted that $\binom{N}{n} = \frac{N!}{n!(N-n)!}$ possible samples of n distinct units of U irrespective of their order of appearances may be identified. Using a Table of Random Numbers, each of such $\binom{N}{n}$ samples from U may be selected with an equal probability $\frac{1}{\binom{N}{n}}$.

A draw-by-draw selection of such a sample may be accomplished by choosing one unit of U in the 1st draw with probability $\frac{1}{N}$, then choosing one of the remaining $(N-1)$ units with probability $\frac{1}{N-1}$ and so on, on the nth draw one of the remaining units $(N - n + 1)$ after thus implementing selection of the first $(n-1)$ distinct units, may be selected with probability $\frac{1}{N-n+1}$. Thus the 1st n-ordered sample $(i_1, i_2, \ldots, i_n)$ of n distinct units of U may be selected with probability $\frac{1}{N(N-1)\ldots(N-n+1)} = \frac{(N-n)!}{N!}$. Thus each of such ordered samples may be selected. Then, the probability of one of unordered samples $\{i_1, i_2, \ldots, i_n\}$, n in numbers may be selected with probability $\frac{(N-n)!n!}{N!} = \frac{1}{\binom{N}{n}}$.

Suppose that we may write $N = nk$ such that n and k are both positive integers greater than 1. Then, $k = \frac{N}{n}$ distinct samples of n distinct units of $U = (1, \ldots, i, \ldots, N)$ may be written down as the following clusters:

$$
\begin{pmatrix} 1 \\ k+1 \\ \vdots \\ 1+(n-1)k \\ 1 \end{pmatrix}
\begin{pmatrix} 2 \\ k+2 \\ \vdots \\ 2+(n-1)k \\ 2 \end{pmatrix}
\cdots
\begin{pmatrix} i \\ k+i \\ \vdots \\ i+(n-1)k \\ i \end{pmatrix}
\cdots
\begin{pmatrix} k \\ 2k \\ \vdots \\ nk \\ k \end{pmatrix}
$$

numbered $1, 2, \ldots, k$.

Then, using a random number table, one may choose at random one of these k clusters. Thus, the probability of selection of one of these possible k samples is $\frac{1}{k}$. Such a sample-selection scheme is known as 'Systematic Sampling Scheme' with an equal probability. More specifically, this is called 'Linear Systematic Sampling' in order to distinguish it from a more general sampling scheme when N cannot have two integer factors as n and k. When, thus, $\frac{N}{n}$ is not an integer, then instead of the above Linear Systematic Sample (LSS), one may select a 'Circular Systematic Sample' (CSS) in the following way.

Define $k = \frac{N}{n}$, which is the integer part of N divided by n.

For example, $N = 23$ and $n = 5$ then $k = \frac{N}{n} = 4$ and 3 is the remainder on dividing N by n.

To select a CSS when $k = \frac{N}{n}$, let us first choose a number R as an equally likely integer between 1 and N. Then, select units $a_j = (R + jk)mod(N)$, $j = 0, 1, \ldots, n - 1$.

In case of LSS, inclusion probability of every unit $i(= 1, \ldots, N)$ is obviously $\frac{1}{k}$. But in CSS, one may note that the total number of possible samples is N because for each choice of an integer R between 1 and N one gets a sample. Thus, one may write down every sample s thus defined. Then, counting the number of samples containing i, say, n_i, the value of π_i is $\frac{n_i}{N}$. It may be noted that the number of units in a CSS, though each distinct, may not be the same and intended size n of a sample may not be realized. Again counting the number n_{ij} of samples s containing i and j of U, we may take $\pi_{ij} = \frac{n_{ij}}{N}$ as the joint inclusion probability of i and j for this CSS scheme. It is worth noting that for both LSS and CSS, the value of π_{ij} is zero for many pairs (i, j).

Now for LSS and CSS also, finding the Horvitz-Thompson estimator is quite easy but an unbiased estimator for its variance may not exist.

Unequal Probability Sampling

Now let us turn to 'unequal' or 'varying' probability sampling. Let it be possible to find certain positive numbers x_i associated with the respective units i of U. These $x_i's$ are called size-measures of i in U. Also, let $X = \sum_{i=1}^{N} x_i$ and $p_i = \frac{x_i}{X}$. Since $\sum_{i=1}^{N} p_i = 1$, $p_i's$ are called the normed size-measures of i in U. A simple method of selection using these $x_i's$ or $p_i's$ is known as sampling with Probability Proportional to Size (PPS) With Replacement (PPSWR). Writing $c_i = x_1 + x_2 + \cdots + x_i$, $i = 1, \cdots, N$ with no loss of generality, multiplying each x_i by 10^k, with k suitably chosen, one may modify each x_i into a positive integer. In the above c_i, we have already done so, and hence, each c_i is a positive integer and so is $c_N = \sum_{i=1}^{N} x_i = X$.

Now using a Table of Random Numbers, one may choose at random an integer R among the numbers 1 and X. Then, if $c_{i-1} < R \leq c_i$, (with $c_0 = 0$), then the unit i may be taken into the sample. So, the selection probability of i into a sample s is $\frac{c_i - c_{i-1}}{X} = \frac{x_i}{X} = p_i$. On repeating this sample-selection procedure independently $n(\geq 1)$ times, one may get a sample s of size n though in s any unit i may appear more than once. Such a sampling procedure is called PPSWR selection procedure in n draws. On each draw of a sampled person, RR data may be gathered by any RR device, some of which have been described above.

Another method of selection of a sample is known as PPSWOR method. This is done in the following manner. As in case of PPSWR, size-measures x_i and normed size-measures $p_i's$ are supposed to be available. On the 1st draw, a unit i_1, say, is drawn from U with a probability p_{i_1}. A second draw is made from U leaving aside i_1 and a unit is chosen with probability $\frac{p_{i_2}}{1-p_{i_1}}$ by the same method as in PPS sampling of a unit using a Table of Random Numbers, as before $x_i's$ are converted into integers. On the 3rd draw, from U leaving aside the units i_1 and i_2 in the 1st two draws, another unit i_3 is chosen by the PPS method with probability $\frac{p_{i_3}}{1-p_{i_1}-p_{i_2}}$. In this way, a PPSWOR sample may be taken in $n(> 2)$ draws from U. Such a sample $s = (i_1, i_2, \ldots, i_n)$ is obviously an ordered sample.

A more general sample-selection procedure is considered by Murthy (1957) in which on the 1st draw a PPS sample is taken as the unit i chosen with probability p_i and a sample s is taken such that $p(s|i)$ is the conditional probability of choosing a sample s of which i is the unit chosen on the 1st draw and the overall sample s is chosen with the probability $p(s) = \sum_{i=1}^{N} p_i p(s|i)$.

Another scheme of sample selection is called 'sampling with a probability proportional to the sum of the size-measures of the $n(\geq 2)$ units in the sample of size n of distinct units, i.e.

$$p(s) = \frac{\sum_{i \in s} x_i}{X \binom{N-1}{n-1}}.$$

In order to draw such a sample, on the 1st draw, a PPS sample is taken as i with probability $p_i = \frac{x_i}{X}$. This selection is then followed by an SRSWOR sample of size $(n-1)$ from the remaining $(N-1)$ units in the population.

A very useful but simple scheme of sampling is given by Rao, Hartley and Cochran (RHC, 1962). Here, as in PPSWR and PPSWOR schemes, the positive and normed size-measures $p_i (0 < p_i < 1, \sum_{i=1}^{N} p_i = 1)$ are supposed to be available. The population $U = (1, 2, \ldots, i, \ldots N)$ is first randomly divided into n disjoint parts or groups consisting respectively of $N_1, \ldots, N_i, \ldots, N_n$ units such that $\sum_n N_i = N$, $\sum_n$ denoting sum over the n groups. To form these groups, first an SRSWOR of N_1 units is taken from U and then successively from the remaining units of U, SRSWORs of $N_2, N_3, \ldots, N_n$ units are disjointly chosen. $N_i's$ are chosen as $\frac{N}{n}$ each, if $\frac{N}{n}$ is an integer. Else, they are taken as

$$N_i = \frac{N}{n} \text{ for } i = 1, 2, \ldots, m \text{ and}$$
$$= \frac{N}{n} + 1 \text{ for } i = m+1, \ldots, n$$

such that $\sum_n N_i = N$. Here m is uniquely available meeting these conditions. Then, for the ith group of N_i units, $p_j (0 < p_j < 1, \sum_i^N p_j = 1)$ values for j in the ith group $(i = 1, \ldots, n)$ are identified and denoted as $p_{ji} (i = 1, \ldots, n; j = 1, \ldots, N_i)$.

Then, from the ith group, one unit jth, say, is selected by PPS method with probability $\frac{p_{ji}}{Q_i}$, taking $Q_i = p_{i1} + p_{i2} + \cdots + p_{iN_i}$, and this is independently done across all the n groups. The resulting sample of n distinct units is RHC's sample.

Brewer and Hanif (1983) and Chaudhuri and Vos (1988) gave numerous methods of sample selection out of which a few are very suitable to use the Horvitz-Thompson (1952) estimator. A few of them ensure each sample to consist of a constant number n of distinct units each and π_i is proportional to p_i and in fact $\pi_i = np_i$ for every i in U. Such sampling schemes are called Inclusion Probability Proportional to Size (IPPS). One convenient to employ in practice is as follows.

On the 1st draw, take a PPS sample, leaving it aside, take one more by PPS from the remaining $(N - 1)$ units and then calculate as follows: the inclusion probabilities of units among the first two draws.

$$\pi_i(2) = p_i + \sum_{\substack{j \neq i = 1}}^{N} p_i \frac{p_j}{(1 - p_i)},$$

$$\pi_{ij}(2) = \frac{p_i p_j}{1 - p_i} + \frac{p_j p_i}{1 - p_j}.$$

Then, leaving aside i and j chosen on the first two draws, take an SRSWOR in $(n - 2)$ draws from the remaining population of $(N - 2)$ units. Then, calculate the inclusion probabilities $\pi_i(n)$ and $\pi_{ij}(n)$ in the overall sample of n units as

$$\pi_i(n) = \pi_i(2) + (1 - \pi_i(2))\frac{n - 2}{N - 2} \quad \text{and}$$

$$\pi_{ij}(n) = \pi_{ij}(2) + \left(\frac{n - 2}{N - 2}\right)(\pi_i(2) + \pi_j(2) - 2\pi_{ij}(2))$$

$$+ \frac{(n - 2)(n - 3)}{(N - 2)(N - 3)}(1 - \pi_i(2) - \pi_j(2) + \pi_{ij}(2)).$$

It is easy to check that $\sum_{i=1}^{N} \pi_i(2) = 2$, $\sum_{j \neq i} \pi_{ij}(2) = \pi_i(2)$, $\sum_{i \neq j} \sum \pi_{ij}(2) = 2$, $\sum_{i=1}^{N} \pi_i(n) = n$, $\sum_{j \neq i} \pi_{ij}(n) = (n - 1)\pi_i(n)$ and $\sum_{i \neq j} \sum \pi_{ij}(n) = n(n - 1)$.

It is an exercise now to show that $\pi_i(n) = np_i$ ensuring that the scheme is really IPPS.

PPSLSS or PPSCSS also may be executed in the following way. Let $x_i's$ be the positive size-measures reduced to integers associated with the units i of U. Let $X = \sum_{i=1}^{N} x_i$ be such that $\frac{X}{n}$ is not an integer. Then, PPSCSS is feasible. Let $c_i = \sum_1^i x_j, i = 1, \ldots, N$ and $c_0 = 0$. Let a random number R between 1 and X be chosen.

Then, taking $k = \left[\frac{X}{n}\right]$ or $k = \left[\frac{X}{n}\right] + 1$, let $a_j = (R + jk) \bmod (X), j = 0, 1, \ldots, (n - 1)$.

If $\frac{X}{n}$ is an integer, then PPSCSS reduces to a PPSLSS.

2.7 Estimation Procedures

In case an SRSWR or an SRSWOR is selected and RR data are gathered from the sampled units, then the sample mean of the $r_i's$ or $r_i(k)'s$ which have $E_R(r_i) = y_i = E_R(r_i(k))$ is taken as unbiased estimators for $\overline{Y} = \frac{1}{N}\sum_{i=1}^{N} y_i$, which becomes a population proportion θ in case y is a qualitative characteristic. In case y is a quantitative characteristic, $\overline{Y}$ is the population mean of real numbers though we should judiciously observe the 0 or 1 values of $y_i's$ in the formulae; there is no difference in estimation for θ, $\overline{Y}$ or in estimation of variances of the unbiased estimators for θ or $\overline{Y}$. We have so far illustrated estimation of θ or $\overline{Y}$ by sample mean in case of SRSWR and SRSWOR and by the Horvitz-Thompson estimator for varying probability sampling. But some of the following alternative estimators are also effectively applicable.

2.7.1 Hansen-Hurwitz (1943) Estimator

This is $e_{HH} = \frac{1}{n}\sum_{i=1}^{n} \frac{r_u}{p_u}$ based on a PPSWR sample in n draws when r_u and p_u are the r' or $r_i(k)$ values and p_i value for the sampled unit i chosen on the uth draw, $u = 1, 2, \ldots, n$.

Then, $E(e_{HH}) = E_P E_R \left(\frac{1}{n}\sum_{u=1}^{n} \frac{r_u}{p_u}\right) = E_P \left(\frac{1}{n}\sum_{u=1}^{n} \frac{y_u}{p_u}\right)$, y_u being the y_i value if unit i is chosen on the uth draw with probability p_u. Then, $E(e_{HH}) = \sum_{i=1}^{N} y_i = Y$.

Alternatively, $E(e_{HH}) = E_R\left[E_P\left(\frac{1}{n}\sum_{u=1}^{n} \frac{r_u}{p_u}\right)\right] = E_R\left(\sum_{i=1}^{N} r\right)_i = \sum_{i=1}^{N} y_i = Y$.

Thus, e_{HH} is unbiased for Y and hence $\frac{e_{HH}}{N}$ is unbiased for $\overline{Y}$ or θ.

Now,

$$V(e_{HH}) = E_P V_R(e_{HH}) + V_P E_R(e_{HH})$$

$$= E_P\left(\frac{1}{n^2}\sum_{u=1}^{n} \frac{V_u}{p_u^2}\right) + V_P\left(\frac{1}{n}\sum_{u=1}^{n} \frac{y_u}{p_u}\right)$$

$$= \frac{1}{n}\sum_{i=1}^{N} \frac{V_i}{p_i} + \frac{1}{n}\left[\sum_{i=1}^{N} \frac{y_i^2}{p_i} - Y^2\right]$$

$$= \frac{1}{n}\sum_{i=1}^{N} \frac{V_i}{p_i} + \frac{1}{n}\sum_{i<j}^{N}\sum_{=1}^{N} p_i p_j\left(\frac{y_i}{p_i} - \frac{y_j}{p_j}\right)^2.$$

An unbiased estimator for $V(e_{HH})$ is

$$v(e_{HH}) = \frac{1}{n}\sum_{u=1}^{n} \frac{v_u}{p_u} + \frac{1}{2n^2(n-1)}\sum_{u<u'}^{n}\sum_{=1}^{n}\left(\frac{r_u}{p_u} - \frac{r_{u'}}{p_{u'}}\right)^2.$$

with obvious notations.

2.7.2 Rao et al. (1962) Estimator

Rao, Hartley and Cochran's (RHC, 1962) estimator for Y from RR data gathered from an RHC sample.

Letting $r_i's$ (or $r_i(k)$'s) obtained from an RHC sample of n units, an unbiased estimator for Y is $e_{RHC} = \sum_n r_i \frac{Q_i}{p_i}$, denoting by p_i the normed size-measure of a unit chosen from the ith group formed by RHC sampling and Q_i, the sum of the N_i normed size-measures of the units falling in the ith group.

$$E(e_{RHC}) = E_P E_R(e_{RHC}) = E_P \left(\sum_{i=1}^{n} y_i \frac{Q_i}{p_i} \right) = Y$$

$$E(e_{RHC}) = E_R E_P(e_{RHC}) = E_R \left(\sum_{i=1}^{N} r_i \right) = Y.$$

$$V(e_{RHC}) = E_P \left(\sum_{i=1}^{n} V_i \left(\frac{Q_i}{p_i} \right)^2 \right) + V_P(E_R(e_{RHC}))$$

$$= \sum_{i=1}^{N} V_i \frac{Q_i}{p_i} + V_P(t_{RHC}), \quad \text{where } t_{RHC} = \sum_{i=1}^{n} y_i \frac{Q_i}{p_i}$$

$$= \sum_{i=1}^{N} V_i \frac{Q_i}{p_i} + \left(\frac{\sum_{i=1}^{n} N_i^2 - N}{N(N-1)} \right) \sum_{i=1}^{N} \sum_{\substack{j=1 \\ i<j}}^{N} p_i p_j \left(\frac{y_i}{p_i} - \frac{y_j}{p_j} \right)^2.$$

An unbiased estimator of this $V(e_{RHC})$ is

$$v(e_{RHC}) = \sum_{i=1}^{n} v_i \frac{Q_i}{p_i} + \left(\frac{\sum_{i=1}^{n} N_i^2 - N}{N^2 - \sum_{i=1}^{n} N_i^2} \right) \sum_n \sum_n Q_i Q_j \left(\frac{r_i}{p_i} - \frac{r_j}{p_j} \right)^2.$$

2.7.3 Des Raj (1956) Estimator from a PPSWOR Sample

Let $(i_1, i_2, \ldots, i_j, \ldots, i_n)$ be an ordered PPSWOR sample s of size n from U chosen with the probability $p(s) = p_{i_1} \frac{p_{i_2}}{(1-p_{i_1})} \frac{p_{i_3}}{(1-p_{i_1}-p_{i_2})} \cdots \frac{p_{i_n}}{(1-p_{i_1}-p_{i_2}-\cdots-p_{i_{n-1}})}$.

Let $e_1 = \frac{r_{i_1}}{p_{i_1}}$, $e_2 = r_{i_1} + \frac{r_{i_2}}{p_{i_2}(1-p_{i_1})}$, $\ldots$, $e_n = r_{i_1} + r_{i_2} + \cdots + r_{i_{n-1}} + \frac{r_{i_n}}{\frac{p_{i_n}}{(1-p_{i_1}-\cdots-p_{i_{n-1}})}}$

where $E_R(r_i) = y_i$.

Then, $e_D = \frac{1}{n}(e_1 + \cdots + e_n)$ is an unbiased estimator for Y and $e_i's$ are pair-wise uncorrelated so that $V(e_D) = \frac{1}{n^2} \sum_{j=1}^{n} V(e_j)$.

So, $v(e_D) = \frac{1}{2n^2(n-1)} \sum_{j<j'}^{n} \sum (e_j - e_{j'})^2$ is an unbiased estimator of $V(e_D)$.

A formula for $V(e_D)$ in a compact form is not easy to derive though Roychaudhury (1957) has given one in a compact form we decide not to present here.

Corresponding to the ordered PPSWOR sample $s = (i_1, \ldots, i_n)$ above, denoting by s^*, the unordered sample as the set of all the $n!$ samples like s above obtainable on permuting the labels $i_1, \ldots, i_n$ in s, one may consider the symmetrised Des Raj estimator based on s^* defined by

$$e^*(\text{SD}) = \frac{\sum_{s \ni s^*} p(s)e_D(s)}{\sum_{s \ni s^*} p(s)},$$

denoting by $\sum_{s \ni s^*}$ the summation over the samples like s obtained on permuting all the labels in s. Consulting Chaudhuri (2010, p. 19) and Chaudhuri and Samaddar (2022), it is possible to work out

$V(e^*(SD)) = V(e_D) - E(e_D - e^*(SD))^2.$

An unbiased estimator of $V(e^*(SD))$ is $v(e^*) = v(e_D) - (e_D - e^*(SD))^2.$

2.7.4 Murthy's (1957) Estimator

Consulting Chaudhuri (2014, p. 64), one may estimate Y using Murthy's (1957) unbiased estimator

$$e_M = \frac{1}{p(s)} \sum_{i \in s} r_i p(s|i)$$

having variance $V(e_M) = \sum_{i=1}^{N} V_i + \sum_{i}^{N} \sum_{<j}^{N} p_i p_j (\frac{y_i}{p_i} - \frac{y_j}{p_j})^2 (1 - \sum_{s \ni i,j} \frac{p(s|i)p(s|j)}{p(s)})$ of which an unbiased estimator is

$$v(e_M) = \sum_{i \in s} v_i \frac{p(s|i)}{p(s)} + \sum_{i<j\in s} \sum \frac{p_i p_j}{p^2(s)} \left(\frac{r_i}{p_i} - \frac{r_j}{p_j}\right)^2$$
$$[p(s)p(s(i, j)) - p(s|i)p(s|j)].$$

A Special Case: PPSS

For the 'Probability Proportional to Sample Sum of Sizes' (PPSS) scheme, let us recall that on the 1st draw a unit i of U is chosen with p_i, as its selection probability employing PPS scheme and next leaving i aside from the remaining $(N - 1)$ units of U an SRSWOR of $(n - 1)$ units is chosen. Then, for this PPSS scheme in n draws, the overall sample s has selection probability $p(s) = \frac{\sum_{i \in s} x_i}{X} \frac{1}{\binom{N-1}{n-1}} =$

$\sum_{i \in s} \dfrac{p_i}{\binom{N-1}{n-1}}$; also $p(s|i) = \dfrac{1}{\binom{N-1}{n-1}} = p(s|j)$ and $p(s|i,j) = \dfrac{1}{\binom{N-2}{n-2}}$.

Moreover, the ratio estimator $e_R = X \dfrac{\sum_{i \in s} r_i}{\sum_{i \in s} x_i}$ is unbiased for Y, if RRs are gathered from s and $r_i's$ are yielded by one of the RR devices. Then, utilizing the results for Murthy's (1957) e_M, $V(e_M)$ and $v(e_M)$ (defined above), it clearly follows that

$$e_R = \frac{1}{p(s)} \sum_{i \in s} r_i \, p(s|i),$$

$$V(e_R) = \sum_{i=1}^{N} V_i + \sum_{i<j}^{N} \sum_{=1}^{N} p_i p_j \left(\frac{y_i}{p_i} - \frac{y_j}{p_j} \right)^2 \left(1 - \frac{X}{\binom{N-1}{n-1}} \sum_{s \ni i,j} \left(\frac{1}{\sum_{i \in s} x_i} \right) \right)$$

and

$$v(e_R) = \sum_{i \in s} v_i \frac{p(s|i)}{p(s)} + \sum_{i<} \sum_{j \in s} p_i p_j \left(\frac{r_i}{p_i} - \frac{r_j}{p_j} \right)^2$$
$$\left[\left(\frac{X}{\sum_{i \in s} x_i} \right) \left(\frac{N-1}{n-1} - \frac{X}{\sum_{i \in s} x_i} \right) \right].$$

(vide Chaudhuri, 2014, p. 65).

2.7.5 Horvitz-Thompson (1952) Estimator

It is interesting to note the forms of the variance of the Horvitz-Thompson (1952) estimator for Y in a DR survey when the sampling design or scheme is such that in a sample, the units are not all distinct and the number of distinct units in every sample is not the same. To observe this, it is important to note the following consistency conditions among π_i's and π_{ij}'s in general.

$$\sum_{i=1}^{N} \pi_i = \sum_{i=1}^{N} \left(\sum_s p(s) I_{si} \right), \text{ writing}$$

$$I_{si} = \begin{cases} 1 & \text{if } s \text{ contains } i \\ 0 & \text{otherwise} \end{cases}.$$

Then,

$$\sum_{i=1}^{N} \pi_i = \sum_{s} p(s) \sum_{1}^{N} I_{si} = v, \text{ say}$$

$$= \sum p(s)v(s), \text{ writing } v(s) \text{ as the number of distinct units in } s.$$

So, $\sum_{i=1}^{N} \pi_i = E_P(v(s))$.

Next, $\sum_{\substack{j=1 \\ j \neq i}}^{N} \pi_{ij} = \sum_{j \neq i}^{N}\left[\sum_s p(s)I_{sij}\right]$ writing $I_{sij} = I_{si}I_{sj} = 1$ if $s \ni i, j$ and

$= 0$, else.

So, $\sum_{j \neq i} \pi_{ij} = \sum_s p(s)\left[I_{si}\sum_{j \neq i} I_{sij}\right] = \sum_s p(s)I_{si}(v(s) - 1) = \sum_{s \ni i} p(s)v(s) - \pi_i$

$$\sum_{i \neq j}\sum \pi_{ij} = \sum p(s)v(s)\left(\sum_s I_{si}\right) - \sum \pi_i$$

$$= \sum p(s)v^2(s) - E_P(v(s))$$

$$= E_P(v^2(s)) - E_P(v(s))$$

$$= V_P(v(s)) + v^2 - v$$

$$= V_P(v(s)) + v(v - 1).$$

Using these, Chaudhuri and Pal (2002) obtained the general formulae

$$V_P(t_{\text{HT}}) = \sum_i \sum_{<j} (\pi_i \pi_j - \pi_{ij})\left(\frac{y_i}{\pi_i} - \frac{y_j}{\pi_j}\right)^2 + \sum_{i=1}^{N} \frac{y_i^2}{\pi_i}\alpha_i$$

writing $\alpha_i = \frac{1}{\pi_i}\sum_{s \ni i} v(s)p(s) - v = 1 + \frac{1}{\pi_i}\sum_{\substack{j=1 \\ j \neq i}}^{N} \pi_{ij} - v,$

so, $v(t_{\text{HT}}) = \sum_{i<j}\sum_{\in s}\left(\frac{\pi_i \pi_j - \pi_{ij}}{\pi_{ij}}\right)\left(\frac{y_i}{\pi_i} - \frac{y_j}{\pi_j}\right)^2 + \sum_{i \in s}\frac{y_i^2}{\pi_i^2}\alpha_i$ for a DR survey.

Now, $V(e_{\text{HT}}) = \sum_i \sum_{<j}(\pi_i \pi_j - \pi_{ij})\left(\frac{y_i}{\pi_i} - \frac{y_j}{\pi_j}\right)^2 + \sum_{i=1}^{N}\frac{y_i^2}{\pi_i}\alpha_i + \sum_{i=1}^{N}\frac{V_i}{\pi_i}$ and

$$v(e_{\text{HT}}) = \sum_{i<j}\sum_{\in s}\left(\frac{\pi_i \pi_j - \pi_{ij}}{\pi_{ij}}\right)\left(\frac{r_i}{\pi_i} - \frac{r_j}{\pi_j}\right)^2 + \sum_{i \in s}\frac{r_i^2}{\pi_i^2}\alpha_i + \sum_{i \in s}\frac{v_i}{\pi_i}$$

are corresponding results for the Horvitz-Thompson estimator when RR data are gathered from such a sample in general.

For an SRSWOR in n draws, it is easy to check that $\pi_i = \frac{n}{N} \forall i \in U$ and $\pi_{ij} = \frac{n(n-1)}{N(N-1)} \forall i \neq j$ in U. Also $\alpha_i = 0$.

So, for SRSWOR, the Horvitz-Thompson estimator for mean $\overline{Y}$ is the sample mean $\overline{y}$ and the variance and unbiased estimator of $V(\overline{y})$ are same as for $v(t_{\text{HT}})$.

For SRSWR in n draws, however sample mean is quite different from the Horvitz-Thompson estimator (HTE) based on it.

Here, $\pi_i = 1 - \left(\frac{N-1}{N}\right)^n$, $\pi_{ij} = 1 - 2\left(\frac{N-1}{N}\right)^n + \left(\frac{N-2}{N}\right)^n$ as may be checked by considering the set theoretic result, $A \cap B = [A^c \cup B^c]^c$.

So, $t_{\text{HT}}, e_{\text{HT}}$ may be written down when based on SRSWR in n draws in DR and RR surveys.

Similarly, for PPSWR in n draws, $\pi_i = 1 - (1 - p_i)^n$, $\pi_{ij} = 1 - (1 - p_i)^n - (1 - p_j)^n + \left(1 - p_i - p_j\right)^n$, and hence, one may use $t_{\text{HT}}, e_{\text{HT}}$ based on PPSWR samples in DR and RR surveys.

References

Brewer, K. R., & Hanif, M. (1983). *Sampling with unequal probabilities*. Springer.

Chaudhuri, A. (2010). *Essentials of survey sampling*. Prentice Hall of India.

Chaudhuri, A. (2011). *Randomized response and indirect questioning techniques in surveys*. CRC Press.

Chaudhuri, A. (2014). *Modern survey sampling*. Chapman & Hall, CRC, Taylor & Francis.

Chaudhuri, A. (2020). A review on issues of settling the sample-size in surveys: Two approaches— Equal and varying probability sampling—Crises in sensitive cases. *CSA Bulletin, 72*(1), 7–16.

Chaudhuri, A., & Christofides, T. C. (2013). *Indirect questioning in sample surveys*. Springer.

Chaudhuri, A., Christofides, T. C., & Rao, C. R. (2016). *Handbook of statistics, data gathering, analysis and protection of privacy through randomized response techniques: Qualitative and quantitative human traits* (Vol. 34). Elsevier.

Chaudhuri, A., & Dutta, T. (2018). Determining the size of a sample to take from a finite population. *Statistics and Applications, 16*(1), 37–44.

Chaudhuri, A., & Mukerjee, R. (1988). *Randomized responses: Theory and techniques*. Marcel Dekker.

Chaudhuri, A., & Pal, S. (2002). On certain alternative mean square error estimators in in complex survey sampling. *Journal of Statistical Pannning and Inference, 104*, 363–375.

Chaudhuri, A., & Samaddar, S. (2022). Estimating the population mean using a complex sampling design dependent on an auxiliary variable. *Statistics in Transition New Series, 23*(1), 39–54.

Chaudhuri, A., & Sen, A. (2020). Fixing the sample-size in direct and randomized response surveys. *Journal of Indian Society of Agricultural Statistics, 74*(3), 201–208.

Chaudhuri, A., & Vos, J. W. (1988). *Unified theory and strategies of survey sampling*. North-Holland.

Des, R. (1956). Some estimators in sampling with varying probabilities without replacement. *Journal of American Statistical Association, 51*, 269–284.

Eriksson, S. A. (1973). A new model for randomized response. *International Statistical Review, 41*, 101–113.

Fox, J. A. (2016). *Randomized response and related methods: Surveying sensitive data*. Sage.

Fox, J. A., & Tracy, P. E. (1986). *Randomized response: A method of sensitive surveys*. Sage.

Hansen, M. M., & Hurwitz, W. N. (1943). On the theory of sampling from finite populations. *Annals of Mathematical Statistics, 14,* 333–362.

Horvitz, D. G., & Thompson, D. J. (1952). A generalization of sampling without replacement from a finite universe. *Journal of American Statistical Association, 47,* 663–685.

Kuk, A. Y. (1990). Asking sensitive questions indirectly. *Biometrika, 77*(2), 436–438.

Murthy, M. N. (1957). Ordered and unordered estimators in sampling without replacement. *Sankhya, 18,* 379–390.

Rao, J. N. K., Hartley, H. O., & Cochran, W. C. (1962). On a simple procedure of unequal probability sampling without replacement. *Journal of Royal Statistical Society, Series B, 24,* 482–491.

Roychaudhury, D. K. (1957). Unbiased sampling design using information provided by linear function of auxiliary variate. In *Chapter 5, Thesis for Associateship of Indian Statistical Institute, Kolkata.*

Warner, S. L. (1965). Randomized response: A survey technique for eliminating evasive answer bias. *Journal of American Statistical Association, 60,* 63–69.

Chapter 3
How to Use Randomized Response Survey Data Obtained by a Specific Procedure to Judge Its Efficiency Relative to a Possible Rival

3.1 Introduction

Cochran (1953, 1963, 1977) gave a method to examine how stratified SRSWOR may outperform unstratified SRSWOR. Rao (1961) gave an alternative simpler way to do the same. Chaudhuri and Pal (2022) and Chaudhuri and Samaddar (2022) applied Rao's (1961) method to assess the efficacy of a sampling strategy using the survey data gathered by this over a rival strategy that may be contemplated, but not actually implemented. Chaudhuri and Samaddar (2022) gave a few illustrations of sampling strategies in contention.

The present chapter is an attempt to extend Rao's (1961) approach to cover competing sample-selection procedures, RR data gathering techniques (RRT) and use of several unbiased estimators of finite population totals and means. Section 3.2 describes briefly the theoretical procedures for estimating totals and their variances. Section 3.3 covers how to simulate data in order to present numerical findings to compare the relative efficiencies of competing estimation procedures. Section 3.4 includes tables showing the findings and the conclusions to be drawn.

3.2 Estimators Under RRT Sampling Procedures

Sometimes situations emerge in human surveys that necessitate the inclusion of sensitive, stigmatizing and incriminating issues. Investigators may consider it a delicacy to directly ask sampled persons such questions relating, for examples, to habitual drunkenness, unlawfully speedy driving, conjugal misbehaviours, expenses with induced abortions, fraudulent tax-evasions, false claims for public grants-in-aid, experiences in treatment for AIDS and expenses incurred for treatment of venereal diseases and similar items. Even if asked, truthful responses are hard to come by.

A. Chaudhuri et al., *Randomized Response Techniques*,
https://doi.org/10.1007/978-981-99-9669-8_3

Warner (1965) gave his randomized response technique (RRT) as a globally admired procedure. Many ramifications followed. A few are narrated in brief as follows.

Warner's RRT

To implement Warner's RRT, a respondent is approached by an enquirer with a box containing several identical cards marked A and its complement A^c. The respondent is asked to respond with a truthful "yes" or "no" on randomly drawing a card from the box about whether the card marked A or A^c drawn 'matches' or 'does not match' his/her own characteristic A or A^c and returning the card to the box. The enquirer is not allowed to see the selected card type. Hence, the respondent believes not to have revealed his/her true feature and thus confidentiality or secrecy is protected.

Let $U = (1, 2, \ldots, i, \ldots N)$ be the finite survey population of N individuals and $\underline{Y} = (y_1, y_2, \ldots, y_i, \ldots y_N)$ be the vector of values y_i for the ith ($i \in U$) population unit on a real variable y so that

$$y_i = \begin{cases} 1 \text{ if } i \text{ bears } A \\ 0 \text{ if } i \text{ bears } A^c \end{cases} ; \quad i = 1, 2, \ldots, N.$$

Furthermore, we denote the y-value for a unit chosen on the kth draw as $y_k = 1$ or 0 and the randomized response from that unit as $I_k = 1$ or 0 if the units are chosen 'with replacement'.

$$I_k = \begin{cases} 1 & \begin{array}{l} \text{if } k \text{ finds a 'match' in his/her} \\ \text{feature } A \text{ or } A^c \text{ with the card-type} \end{array} . \\ 0 \text{ if there is 'no match'} \end{cases}$$

Also, let us denote $\underline{X} = (x_1, x_2, \ldots, x_i \ldots, x_N)$ as a vector of N coordinates x_i for the units i of U on another variable x such that $0 < x_i < 1$ for $i = 1, 2, \ldots, N$, and $X = \sum_{i=1}^{N} x_i$. These $x_i's$ are called 'size-measures'. Then, $p_i = \frac{x_i}{X}$ for $i = 1, 2 \ldots N$, are called 'normed size-measures'.

Now suppose our intention is to estimate the population total $Y = \sum_{i=1}^{N} y_i$, or the population mean $\overline{Y} = \frac{Y}{N}$, and for this, a sample s of size $n(2 \leq n \leq N)$ is drawn from U by the 'Probability Proportional to Size With Replacement' (PPSWR) method using the values of $\underline{X}$ which is fixed and known. The values y_k of course are unknown and cannot be ascertained. But I_k values are gathered with respective probabilities $p_k = \frac{x_k}{X}$ in n independent draws.

Let us denote E_R, V_R as expectation and variance operators with respect to any RR method.

Then, for Warner's RRT with a proportion p of cards marked A and $(1 - p)$ of cards marked A^c so that $0 < p < 1$ and $p \neq \frac{1}{2}$

$$E_R(I_k) = py_k + (1 - p)(1 - y_k).$$

Also,

$$V_R(I_k) = p(1 - p).$$

Then, $r_k = \frac{I_k - (1-p)}{(2p-1)}$ has $E_R(r_k) = y_k$.

Also, $V_R(r_k) = \frac{1}{(2p-1)^2} V_R(I_k) = \frac{p(1-p)}{(2p-1)^2}$ as $I_k^2 = I_k$ and $y_k^2 = y_k$.

Using the I_k and V_k values from a PPSWR sample s of size n, we get

$$e_{HH} = \frac{1}{n} \sum_{i=1}^{N} \frac{r_i}{p_i} f_i$$

for which $E_R(e_{HH}) = \frac{1}{n} \sum_{k=1}^{n} \frac{y_k}{p_k}$, $V_R(e_{HH}) = \frac{1}{n} V_R\left(\frac{r_1}{p_1}\right)$ and $E_P V_R(e_{HH}) = \frac{1}{n} \frac{p(1-p)}{(2p-1)^2} \sum_{i=1}^{N} \frac{1}{p_i}$. Here, f_i is the frequency of i in the PPSWR sample s in n draws.

Now denoting E_P, V_P as the operators generically design-based expectation and variance, and by $E = E_P E_R = E_R E_P$, $V = E_P V_R + V_P E_R = E_R V_P + V_R E_P$ as the overall design and RR-based expectation, variance operators, each generically, we get

$$E(e_{HH}) = Y \text{ and}$$

$$V(e_{HH}) = E_P E_R (e_{HH} - Y)^2 = E_R E_P (e_{HH} - Y)^2$$

$$= \frac{1}{n} \sum_{i=1}^{N} p_i \left(\frac{y_i}{p_i} - Y\right)^2$$

$$+ \left(\frac{1}{n} \frac{p(1-p)}{(2p-1)^2}\right) \sum_{i=1}^{N} \frac{1}{p_i} + \frac{n-1}{n} N \left(\frac{p(1-p)}{(2p-1)^2}\right).$$

It follows that

$$v(e_{HH}) = \frac{1}{n(n-1)} \sum_{i=1}^{N} f_i \left(\frac{r_i}{p_i} - e_{HH}\right)^2$$

$$+ \left(\frac{1}{n} \frac{p(1-p)}{(2p-1)^2}\right) \sum_{i=1}^{N} \frac{f_i}{p_i} \text{ has}$$

$$E(v(e_{HH})) = E_P E_R v(e_{HH}) = E_R E_P v(e_{HH}) = V(e_{HH}),$$

i.e. $v(e_{HH})$ is an unbiased estimator of $V(e_{HH})$, the variance of the Hansen-Hurwitz (HH) estimator e_{HH} for Y. The above expressions are due to Chaudhuri et al. (2016, Chap. 12).

3.2.1 PPSWR Sampling with Hansen-Hurwitz Estimator Versus SRSWR with Expansion Estimator

In case a PPSWR sample s of size n is drawn from U and RR data are obtained then by Warner's (1965) RR with the PPSWR sampling and combining this with Hansen-Hurwitz estimator is more complicated than SRSWR sampling. However, combining that with the expansion estimator coupled with RRTs would be a simpler alternative.

3.2.1.1 Warner's RRT

PPSWR reduces to SRSWR when $p_i = \frac{1}{N} \forall i$ in U. Therefore, the above formulas become:

$$e_{\text{SRSWR,RR}} = \frac{N}{n} \sum_{k=1}^{n} r_k$$

then variance would have been

$$V\left(e_{\text{SRSWR,RR}}\right) = \frac{N}{n} \sum_{i=1}^{N} (y_i - \overline{Y})^2 + \frac{N^2}{n} \left(\frac{p(1-p)}{(2p-1)^2} \right).$$

Given the Warner-based RR survey data at hand gathered by PPSWR sampling in n draws, an unbiased estimate of this variance would be derived as follows:

$$\widehat{Y^2} = e_{\text{HH}}^2 - v(e_{\text{HH}}) = \left(\sum_{i=1}^{N} \frac{r_i f_i}{n p_i} \right)^2 - v(e_{\text{HH}}).$$

So, an unbiased estimate of $V\left(e_{\text{SRSWR,RR}}\right)$ would be writing $e_{\text{HH}}(Y^2)$ based on y_k^2's though in this qualitative case $y_k^2 = y_k$ for all k,

$$v\left(e_{\text{SRSWR,RR}}\right) = \frac{N}{n} \left(\sum_{i=1}^{N} \widehat{y_i^2} - N\widehat{\overline{Y}^2} \right) + \frac{N^2}{n} \left(\frac{p(1-p)}{(2p-1)^2} \right)$$

$$= \frac{N}{n} \left(\sum_{i=1}^{N} \widehat{y_i^2} - \frac{\widehat{Y^2}}{N} \right) + \frac{N^2}{n} \left(\frac{p(1-p)}{(2p-1)^2} \right)$$

$$= \frac{N}{n} \left(e_{\text{HH}}(Y^2) - \frac{\left(\sum_{i=1}^{N} \frac{r_i f_i}{n p_i} \right)^2 - v(e_{\text{HH}})}{N} \right) + \frac{N^2}{n} \left(\frac{p(1-p)}{(2p-1)^2} \right).$$

Now, based on Warner's RR model, gain in efficiency of PPSWR using Hansen-Hurwitz estimator over a possible alternative of SRSWR, expansion estimate is

$$G_{RR} = v(e_{SRSWR,RR}) - v(e_{HH})$$

$$= \frac{1}{n}\left(N\sum_{i=1}^{N}\frac{r_i f_i}{np_i} - \left(\sum_{i=1}^{N}\frac{r_i f_i}{np_i}\right)^2\right)$$

$$+ \frac{N^2}{n}\left(\frac{p(1-p)}{(2p-1)^2}\right) - \left(1-\frac{1}{n}\right)v(e_{HH}).$$

In this qualitative case with every y_i as either 1 or 0, we may also permit the following other RRTs.

3.2.1.2 URL [or Unrelated Features Given by Simmons' (Vide Greenberg et al. (1969) and Horvitz et al. (1967))] RRT

Presuming A^c, the complement of A also stigmatizing, say A denoting supporting a particular party and A^c a rival party, Simmons recommended another RR device where an enquirer approaches a sampled person with two boxes of identical cards, in one of them with A -marked cards in proportion $p_1 (0 < p_1 < 1)$ and the remaining ones marked B in proportion $(1 - p_1)$ taking B as a feature unrelated to A and innocuous in nature like preferring music to painting and in the other box A-marked and B- marked cards are in proportions $p_2: (1 - p_2)$, $p_1 \neq p_2$, $(0 < p_2 < 1)$.

Here we write

$$y_i = \begin{cases} 1 \text{ if } i \text{ bears } A \\ 0 \text{ if } i \text{ bears } A^c \end{cases} \text{ and } x_i = \begin{cases} 1 \text{ if } i \text{ bears } B \\ 0 \text{ if } i \text{ does not bear } B \end{cases}.$$

and

$$I_i = \begin{cases} 1 \text{ if card type drawn from 1st box 'matches' } A \text{ or } B \\ 0 \text{ if 'no match' with } i's \text{ actual fact} \end{cases}$$

$$J_i = \begin{cases} 1 \text{ if card type from 2nd box 'matches'} \\ 0 \text{ if does not} \end{cases}.$$

Then, $E_R(I_i) = p_1 y_i + (1 - p_1)x_i$ and $E_R(J_i) = p_2 y_i + (1 - p_2)x_i$.

$$r_i = \frac{(1 - p_2)I_i - (1 - p_1)J_i}{(p_1 - p_2)}.$$

$$E_R(r_i) = y_i, \ y_i^2 = y_i, \ x_i^2 = x_i, \ I_i^2 = I_i, \ J_i^2 = J_i.$$
$$V_R(I_i) = p_1(1 - p_1)(y_i - x_i)^2$$

$$V_R(J_i) = p_2(1 - p_2)(y_i - x_i)^2$$

$$V_R(r_i) = \frac{(1 - p_1)(1 - p_2)(p_1 + p_2 - 2p_1 p_2)}{(p_1 - p_2)^2}(y_i - x_i)^2.$$

Since

$$\begin{aligned} V_i = V_R(r_i) &= E_R\left(r_i^2\right) - (E_R(r_i))^2 \\ &= E_R\left(r_i^2\right) - y_i^2 = E_R\left(r_i^2\right) - y_i \\ &= E_R\left(r_i^2 - r_i\right) = E_R(r_i(r_i - 1)), \end{aligned}$$

so, $v_i = r_i(r_i - 1)$ is an unbiased estimator of V_i. Because $r_i(r_i - 1)$ may turn out negative for many $i's$, it may not be a realistic variance estimator in an estimator for the variance of an unbiased estimator for a finite population total. This negativity of v_i often is not consequential as we shall note in our illustration.

3.2.1.3   Kuk (1990)'s RRT

Here, the interviewer approaches the respondent with two boxes of identical cards with proportions of "red" versus "non-red" coloured cards as θ_1 in the 1st box and θ_2 in the 2nd, $(\theta_1, \theta_2$ as $0 < \theta_i < 1, i = 1, 2$ but $\theta_1 \neq \theta_2)$. If the addressee bearing A is to use the 1st box and if he/she bears A^c is to use the 2nd box and draw cards at random with replacement $k(> 1)$ times and the ith respondent is to announce the number $f_i(0 \le f_i \le k \, \forall i \in U)$ as his/her RR, it follows then

$$\begin{aligned} E_R(f_i) &= k[y_i\theta_1 + (1 - y_i)\theta_2] \text{ and} \\ V_R(f_i) &= k[y_i\theta_1(1 - \theta_1) + (1 - y_i)\theta_2(1 - \theta_2)]. \end{aligned}$$

Consequently, $r_i(k) = \frac{f_i/k - \theta_2}{\theta_1 - \theta_2}$ has $E_R(r_i(k)) = f_i$ and $V_R(r_i(k)) = \frac{V_R(f_i/k)}{(\theta_1 - \theta_2)^2} = V_i(k)$ and an unbiased estimator for $V_i(k)$ is $v_i(k) = b_i(k)r_i(k) + c_i(k)$, writing $b_i(k) = \frac{1 - \theta_1 - \theta_2}{k(\theta_1 - \theta_2)^2}$ and $c_i(k) = \frac{\theta_2(1 - \theta_2)}{k^2(\theta_1 - \theta_2)^2}$.

3.2.1.4   Christofides (2003)'s RRT

The inquirer is to approach a sampled person with a box of $M(> 2)$ cards identical but marked $1, 2, \ldots, k, \ldots M$ occurring in proportion p_k with $0 < p_k < 1$, $\sum_1^M p_k = 1$ and on request obtain the RR as $(M + 1 - k)$ if his/her feature is A and he/she randomly chooses the card marked k if his/her feature is A^c then from him/her the RR is k if he/she randomly chooses the card marked k. Then from the ith person, the RR is

$$z_i = (M + 1 - k)y_i + k(1 - y_i), i \in U \text{ and } k = 1, 2, \ldots, M$$

Then, $E_R(k) = \sum_1^M kp_k = \mu$ and $V_R(k) = \sum_1^M k^2 p_k - \mu^2$ and so,
$E_R(z_i) = y_i(M + 1 - 2\mu) + \mu = \mu_i$, say and
$V_R(z_i) = E_R(z_i^2) - \mu_i^2 = E_R[(M + 1 - k)^2 y_i + k^2(1 - y_i)] - \mu_i^2$ because $y_i^2 = y_i \forall i \in U$,

$$
V_R(z_i) = \sum_{k=1}^M p_k k^2 + y_i \sum_{k=1}^M p_k[(M + 1 - k)^2 - k^2] - \mu_i^2
$$

$$
= \sum_{k=1}^M p_k k^2 + y_i \sum_{k=1}^M p_k[(M + 1)(M + 1 - 2k)]
$$
$$
- \mu^2 - y_i(M + 1 - 2\mu)^2 - 2\mu y_i(M + 1 - 2\mu)
$$

$$
= \sum_{k=1}^M p_k k^2 - \mu^2 + y_i[(M + 1)^2 - 2(M + 1)\mu
$$
$$
- (M + 1)^2 - 4u^2 + 4(M + 1)\mu - 2(M + 1)\mu + 4\mu^2]
$$

$$
= \sum_1^M p_k k^2 - \mu^2 = V_R(k)
$$

Since M and μ are known, we may take $r_i = \frac{z_i - \mu}{M + 1 - 2\mu}$ to get $E_R(r_i) = y_i$ and
$V_R(r_i) = \frac{V_R(z_i)}{(M + 1 - 2\mu)^2} = V$, say.

3.2.1.5 Boruch's (1972) Forced Response RRT

The interviewer approaches a sampled person with a number of similar looking cards in proportions $p_1, p_2, 1 - p_1 - p_2$ each positive and less than 1 and respectively marked "yes", "no" and "genuine" and is asked randomly to choose 1 card from the box and report respectively "yes", "no" and genuinely say yes or no accordingly as it bears A or A^c.

Then, from the sampled person i, the RR is $I_i = \begin{cases} 1 \text{ if } i \text{ responds "Yes"} \\ 0 \text{ if } i \text{ responds "No"} \end{cases}$.

Then,

$$
P(I_i = 1 | y_i = 1) = p_1 + (1 - p_1 - p_2) = 1 - p_2
$$

$$
P(I_i = 0 | y_i = 1) = p_2
$$

$$
P(I_i = 0 | y_i = 0) = p_2 + (1 - p_1 - p_2) = 1 - p_1
$$

$$
P(I_i = 1 | y_i = 0) = p_1.
$$

So, $E_R(I_i) = P(I_i = 1) = p_1 + (1 - p_1 - p_2)y_i$.

So, letting $r_i = \frac{I_i - p_1}{(1 - p_1 - p_2)}$, $E_R(r_i) = y_i$ and $V_R(r_i) = V_i = \frac{V_R(I_i)}{(1 - p_1 - p_2)^2} = \frac{p_1(1 - p_1) + y_i(1 - p_1 - p_2)(p_2 - p_1)}{(1 - p_1 - p_2)^2}$.

3.2.1.6 Mangat and Singh (1990)'s RRT

The interviewer approaches a selected person i with two boxes. The 1st box contains a proportion $T (0 < T < 1)$ of cards marked 'true' and the remaining ones marked 'random'. On request, the respondent is to draw randomly one card and if a 'true' marked is drawn, he/she is to give out the true value y_i which equals 1 or 0, and if a 'randomized' card is drawn, he/she is to randomly draw a card from the other box. This contains a proportion $p(0 < p < 1)$ and $p \neq \frac{1}{2}$ of cards marked A, and the others marked A^c and randomly drawing one card is to report

$$I_i = \begin{cases} 1 \text{ if } i's \text{ card type drawn matches his actual feature } A \text{ or } A^c \\ 0 \text{ if it mismatches} \end{cases}$$

and

$$z_i = \begin{cases} y_i \text{ with probability } T \\ 0 \text{ with probability } (1 - T) \end{cases}.$$

For this, RRT, the RR for i is z_i with

$$E_R(z_i) = T y_i + (1 - T)[p y_i + (1 - p)(1 - y_i)]$$

or $E_R(z_i) = (1 - T)(1 - p) + y_i[T + (1 - T)(2p - 1)] = \alpha + y_i(1 - 2\alpha)$; taking $\alpha = (1 - T)(1 - p)$.

So, $r_i = \frac{z_i - (1 - T)(1 - p)}{T + (1 - T)(2p - 1)}$, provided $T + (1 - T)(2p - 1) \neq 0$ iff $T \neq \frac{1 - 2p}{2(1 - p)}$.

Then, $V_R(z_i) = E_R(z_i)(1 - E_R(z_i))$ because $z_i^2 = z_i$. On simplifying $V_R(z_i) = [\alpha + y_i(1 - 2\alpha)][(1 - \alpha) - y_i(1 - 2\alpha)] = \alpha(1 - \alpha)$ because $y_i^2 = y_i$.

So, $V_R(z_i) = (1 - T)(1 - p)(T + p(1 - T))$.

So, $V = V_R(r_i) = \frac{(1 - T)(1 - p)[T + p(1 - T)]}{(T + (1 - T)(2p - 1))^2}$.

If $T = 0$, V equals $\frac{p(1 - p)}{(2p - 1)^2}$ as it should.

There are many such RRTs in the literature concerning qualitative characteristics A, A^c, B, B^c.

Quantitative Randomized Response

We shall illustrate only two randomized response devices concerning quantitative features. Here, y is a real variable, and for an individual i, the variable takes any real value $y_i, i \in U$. The 1st RR procedure is (vide Chaudhuri, 2011) Device I.

3.2.1.7 Device I

The inquirer approaches a sampled person i with two boxes. One box contains identical cards marked $a_1, a_2, \ldots, a_T$ and a second box contains cards marked $b_1, b_2, \ldots, b_m$, with T and m are quite large, and these numbers a_j and b_k are all distinct with means and variances $\mu = \frac{1}{T} \sum_{j=1}^{T} a_j \neq 0, \sigma_A^2 = \frac{1}{T} \sum_{j=1}^{T} (a_j - \mu)^2$ and $\gamma = \frac{1}{m} \sum_{k=1}^{m} b_k, \sigma_B^2 = \frac{1}{m} \sum_{k=1}^{m} (b_k - \gamma)^2$. The sampled person i on request is to draw one card each from the 1st and the 2nd boxes, say, labelled a_j and b_k and give out the RR as $z_i = a_j y_i + b_k$.

Then, $E_R(z_i) = \mu y_i + \gamma$ and $V_R(z_i) = \sigma_A^2 y_i^2 + \sigma_B^2$.

Then, $r_i = \frac{z_i - \gamma}{\mu}, E_R(r_i) = y_i$ and $V_R(r_i) = \left(\frac{\sigma_A^2}{\mu^2}\right) y_i^2 + \left(\frac{\sigma_B^2}{\mu^2}\right) = \alpha y_i^2 + \beta, say.$

Hence, $E_R(r_i^2) = (1 + \alpha) y_i^2 + \beta.$

Then, $v_R(r_i) = \frac{\alpha r_i^2 + \beta}{1 + \alpha}, E_R(v_R(r_i)) = V_R(r_i).$

It should be noted that $E_R(r_i^2 - v_R(r_i)) = y_i^2$. In other words, $r_i^2 - v_R(r_i)$ is an unbiased estimator of y_i^2 for quantitative cases.

3.2.1.8 Device II (Vide Chaudhuri, 2011; Eriksson, 1973)

Here, the inquirer offers a sampled person i a box with a large number of cards, a proportion $C(0 < C < 1)$ of them marked 'correct' and the remaining numerous cards marked $x_1, x_2, \ldots, x_m$ in proportions $q_1, q_2, \ldots, q_m$ such that $\sum_{j=1}^{m} q_j = 1 - C$. The sampled person i on request is to randomly choose one card, and to give out the RR as y_i if a 'correct', marked card is chosen or the RR as x_j if drawn.

Thus, $z_i = \begin{cases} y_i & \text{with probability } C \\ x_j & \text{with probability } q_j \end{cases}$.

Then, $E_R(z_i) = C y_i + \sum_{j=1}^{m} q_j x_j.$

Taking $r_i = \frac{1}{C}\left(z_i - \sum_{j=1}^{m} q_j x_j\right)$, one gets $E_R(r_i) = y_i$ and

$V_i = V_R(r_i) = \frac{1}{C^2} V_R(z_i) = \frac{1}{C^2}\left(E_R(z_i^2) - (E_R(z_i))^2\right) = \alpha y_i^2 + \beta y_i + \varphi$, by Eriksson (1973).

These α, β and φ are the collecting terms involving y_i^2, y_i and those free of $y_i's.$

It follows that $v_i = v_R(r_i) = \frac{\alpha r_i^2 + \beta r_i + \varphi}{(1 + \alpha)}$ satisfies $E_R(v_i) = V_i$, thus v_i providing an unbiased estimator for $V_i = V_R(r_i)$.

We have already given our theory for evaluating gain in efficiency in employing (Sect. 3.2.1) PPSWR sampling combined with Hansen-Hurwitz estimator by

Warner's RRT data over SRSWR plus expansion estimator with the same RRT data. For each of the other RRT's qualitative as well as quantitative RRT data, the change is obvious; simply use the correct $v_i's$ already given. Now let us present the same for comparisons.

3.2.2 PPSWOR Sampling with Des Raj Estimator Versus SRSWOR with Expansion Estimator

If a PPSWOR sample of size n is taken, the draw-by-draw selection probabilities are $\frac{p_{ir}}{1 - p_{i1} - p_{i2} - \cdots - p_{i(r-1)}}, r = 1, 2, \ldots, n$ and draw-by-draw unbiased estimators for Y are

$$
\begin{aligned}
e_1 &= \frac{r_{i1}}{p_{i1}}, \\
e_2 &= r_{i1} + \frac{r_{i2}}{p_{i2}}(1 - p_{i1}), \\
&\cdots, \\
e_n &= r_{i1} + r_{i2} + \cdots + r_{i(n-1)} + \frac{r_{in}}{p_{in}}\left(1 - p_{i1} - p_{i2} - \cdots - p_{i(n-1)}\right).
\end{aligned}
\tag{3.1}
$$

Here r_i is an unbiased estimator of y_i, i.e. $E_R(r_i) = y_i$ and $V_R(r_i) = V_i$.
Thus,

$$
e_D = \frac{1}{n}(e_1 + e_2 + \cdots + e_n) = \bar{e} \,(\text{say})
$$

is the Des Raj estimator of Y because $E_P(e_D) = \sum_{i=1}^{N} r_i = R$ (say) and $E(\bar{e}) = E_R E_P(e_D) = E_R(R) = \sum_{i=1}^{N} y_i = Y$.
Its unbiased variance estimator for $V_p(e_D)$ is

$$
v_P(e_D) = \frac{1}{n(n-1)} \sum_{i=1}^{n}(e_i - \bar{e})^2 = \frac{1}{2n^2(n-1)} \sum_{i \neq j}\sum (e_i - e_j)^2
$$

because $E_P\big(v_p(e_D)\big) = V_P(e_D)$.
Note that, $V(e_D) = E_R V_P(e_D) + V_R E_P(e_D) = E_R V_P(e_D) + V_R(R) = E_R V_P(e_D) + \sum_{i=1}^{N} V_i$.
So, its unbiased estimate is $v(e_D) = v_P(e_D) + e_D(v)$. Here, $e_D(v)$ is the Des Raj estimator of $\sum_{i=1}^{N} V_i$, replacing r_{ij} term in (3.1) by $v_{ij} = r_{ij}(r_{ij} - 1)$.
For SRSWOR, unbiased estimator of Y is the expansion estimator which is $N\bar{r}$, with $\bar{r} = \frac{1}{n} \sum_{i \in s} r_i$.

$$
V(N\bar{r}) = V(N\bar{y}) + \frac{N}{n} \sum_{i=1}^{N} V_i
$$

$$= N^2 \left(\frac{N-n}{Nn} \right) \frac{1}{N-1} \sum_{i=1}^{N} (y_i - \bar{Y})^2 + \frac{N}{n} \sum_{i=1}^{N} V_i$$

$$= C_1 \left[\sum_{i=1}^{N} y_i^2 - \frac{Y^2}{N} \right] + \frac{N}{n} \sum_{i=1}^{N} V_i,$$

writing $C_1 = N \left(\frac{N-n}{n} \right) \frac{1}{N-1}$.

Now, $V(e_D) = E(e_D^2) - Y^2$.

So, from PPSWOR in n draws $\widehat{Y^2} = e_D^2 - v(e_D)$.

Also, $\sum_{i=1}^{N} y_i^2$ is unbiasedly estimated by $e_D(y^2)$ which is e_D with each y in e_D replaced by y^2; of course, for qualitative cases, $y^2 = y$ because each y_i equals only 1 or 0. So, unbiased estimate of $V(N\bar{r})$ based on PPSWOR is

$$v(N\bar{r}) = C_1 \left[e_D(y^2) - \frac{\widehat{Y^2}}{N} \right] + \frac{N}{n} \sum_{i=1}^{N} \widehat{V_i}.$$

So, the gain in efficiency is $G_{RR} = v(N\bar{r}) - v(e_D)$ from above.

3.2.3 RHC Sampling and Estimation Versus SRSWOR Expansion Estimator

Based on RHC sample, for $r_i's$ with $E_R(r_i) = y_i$, the RHC estimator for Y is $e_{RHC} = \sum_{i=1}^{n} r_i \frac{Q_i}{p_i}$, $Q_i = \sum_{j=1}^{N_i} P_{ij}$.

$$E_R(e_{RHC}) = \sum_{i=1}^{n} y_i \frac{Q_i}{p_i}$$

$$E(e_{RHC}) = E_P E_R(e_{RHC}) = \sum_{i=1}^{N} y_i = Y$$

$$V(e_{RHC}) = V_P \left(\sum_{i=1}^{n} y_i \frac{Q_i}{p_i} \right) + E_P \left(\sum_{i=1}^{n} V_i \left(\frac{Q_i}{p_i} \right)^2 \right)$$

$$= \left(\frac{\sum_{i=1}^{n} N_i^2 - N}{N(N-1)} \right) \left\{ \sum_{i=1}^{N} p_i \left(\frac{y_i}{p_i} - Y \right)^2 \right\} + \sum_{i=1}^{N} V_i \left(\frac{Q_i}{p_i} \right)$$

$$v(e_{RHC}) = \left(\frac{\sum_{i=1}^{n} N_i^2 - N}{N^2 - \sum_{i=1}^{n} N_i^2} \right) \left\{ \sum_{i=1}^{n} Q_i \left(\frac{r_i}{p_i} \right)^2 - e_{RHC}^2 \right\} + \sum_{i=1}^{n} V_i \frac{Q_i}{p_i}$$

$$V(e_{RHC}) = E(e_{RHC}^2) - Y^2$$

So, $\widehat{Y^2} = e_{\text{RHC}}^2 - v(e_{\text{RHC}})$.

For SRSWOR, unbiased estimator for Y is

$$e_{\text{SRS}} = \frac{N}{n} \sum_{i \in s} r_i$$

and

$$V_{\text{SRSWOR}}(e_{\text{SRS}}) = N^2 \left(\frac{N-n}{Nn} \right) \frac{1}{N-1} \left(\sum_{i=1}^{N} y_i^2 - \frac{Y^2}{N} \right) + \frac{N^2}{n^2} \left(\sum_{i=1}^{N} V_i \right) \frac{n}{N}$$

$$= N^2 \frac{(N-n)}{Nn} \frac{1}{(N-1)} \sum_{i=1}^{N} (y_i - \overline{Y})^2 + \frac{N}{n} \sum_{i=1}^{N} V_i = V.$$

To unbiasedly estimate V from RHC sample,

$$\widehat{Y_{\text{RHC}}^2} = e_{\text{RHC}}^2 - v(e_{\text{RHC}}) = \left(\sum_{i=1}^{n} \frac{r_i Q_i}{p_i} \right)^2 - v(e_{\text{RHC}}).$$

So, $v_{\text{SRSWOR}}(e_{\text{SRS}}) = C_1 \left(e_{\text{RHC}}(y^2) - \frac{\widehat{Y_{\text{RHC}}^2}}{N} \right) + \frac{N}{n} \widehat{\sum_{i=1}^{N} V_i}$, writing $C_1 = N^2 \frac{(N-n)}{Nn} \frac{1}{(N-1)}$.

Here, $e_{\text{RHC}}(y^2)$ is RHC estimate of $\sum_{i=1}^{N} y_i^2$ but $\sum_{i=1}^{N} y_i^2 = \sum_{i=1}^{N} y_i$ because $y_i = 1$ or 0 for qualitative cases.

Now, $G_{\text{RR}} = v_{\text{SRSWOR}}(e_{\text{SRS}}) - v(e_{\text{RHC}})$ is the gain in efficiency.

3.2.4   Lahiri-Midzuno-Sen (LMS) Horvitz-Thompson Estimator Versus SRSWOR Expansion Estimator

Based on LMS sample, the Horvitz-Thompson (HT) estimator for Y is

$$e_{\text{LMS}} = \sum_{i=1}^{n} \frac{r_i}{\pi_i},$$

where $\pi_i = \frac{N-n}{N-1} p_i + \frac{n-1}{N-1}$.

$$v(e_{\text{LMS}}) = \sum_{i,j=1}^{n} \sum_{j>i} \frac{(\pi_i \pi_j - \pi_{ij})}{\pi_{ij}} \left(\frac{r_i}{\pi_i} - \frac{r_j}{\pi_j} \right)^2 + \sum_{i=1}^{n} \frac{v_i}{\pi_i}, \text{ where } \pi_{ij} = \frac{(n-1)(N-n)}{(N-1)(N-2)} (p_i + p_j) + \frac{(n-1)(n-2)}{(N-1)(N-2)}.$$

So, $v_{\text{SRSWOR, LMS}}(e_{\text{SRS}}) = C_1\left(e_{\text{LMS}}(y^2) - \dfrac{\widehat{Y^2_{\text{LMS}}}}{N}\right) + \dfrac{N}{n}\widehat{\sum_{i=1}^{N} V_i}$ unbiasedly estimate V and $v_{\text{SRSWOR,LMS}}(e_{\text{SRS}}) - v(e_{\text{LMS}})$ is the gain in efficiency.

3.2.5 Lahiri-Midzuno-Sen (LMS) Ratio Estimator Versus SRSWOR Expansion Estimator

The ratio estimator for Y is

$$e^*_{\text{LMS}} = \frac{\sum_{i=1}^{n} r_i}{p_s} \quad ; p_s = \sum_{i\epsilon s} p_i$$

and an unbiased variance estimator is

$$v\left(e^*_{\text{LMS}}\right) = \sum_{i<j\epsilon s}\sum\left(\frac{N-1}{n-1}\frac{1}{p_s} - \frac{1}{p_s^2}\right)$$

$$p_i p_j\left[\left(\frac{r_i}{p_i} - \frac{r_j}{p_j}\right)^2 - \left(\frac{v_i}{p_i^2} + \frac{v_j}{p_j^2}\right)\right] + \sum_s \frac{v_i}{p_s^2}.$$

So, an unbiased estimator of V from the LMS sample is

$$v_{\text{SRSWOR, LMS}*}(e_{\text{SRS}}) = C_1\left(e^*_{\text{LMS}}(y^2) - \frac{e^{*2}_{\text{LMS}} - v\left(e^*_{\text{LMS}}\right)}{N}\right) + \frac{N}{n}\widehat{\sum_{i=1}^{N} V_i}.$$

For many aspects of this section, Adhikary (2016) is handy to consider.

3.3 Numerical Computations and Simulation

For both qualitative and quantitative characteristics, we use numerical data borrowed from the comprehensive work by Chaudhuri et al. (2009), Chaudhuri and Christofides (2013).

Here,

$$N = 116, \ Y = \begin{cases} 96 & \text{(qualitative)} \\ 105{,}336 & \text{(quantitative)} \end{cases}.$$

For URL model, an innocuous character is also considered. A size-measure variable is considered here to draw samples in varying probability sampling schemes. For simulation, different sample-sizes with different parameter values are considered. Relative gain in efficiency is shown here using $100 \frac{G_{\text{RR,Design}}}{v(e_{\text{SRSWR(WOR),Design}})}$ along with the actual gain in efficiency ($G_{\text{RR,Design}}$).

In Table 3.1, the details required to evaluate $\sum_{i=1}^{N} V_i$ using RR survey data at hand are presented for various RRTs. For example, Warner's RR model has known $V_i = \frac{p(1-p)}{(2p-1)^2}$. Therefore, for this RR model, $\widehat{\sum_{i=1}^{N} V_i}$ is the sum of all V_i's. However, V_i of Kuk's model depends on the unknown y-value. Thus, $\widehat{\sum_{i=1}^{N} V_i}$ is estimated here from the PPS sample at hand, and the estimator is mentioned in the 3rd column of Table 3.1.

Tables 3.2, 3.3, 3.4, 3.5 and 3.6 present the numerical estimates of the gain in efficiency in employing complex strategies versus simpler alternatives, considering the formulae for various sampling strategies combined with various RRTs. These tables are reconsidered here from the article by the authors of this monograph (Chaudhuri and Patra (2023)).

Different parameter values are taken in the 2nd column of the mentioned tables. For example, the parameter p is taken as $0.3, 0.4$ and 0.7 for Warner's RR model. In Table 3.2 for URL model, parameters (p_1, p_2) are taken as $(0.25, 0.55), (0.25, 0.65)$ and $(0.4, 0.6)$. Estimated population total (e_{Design}) and its unbiased variance estimate ($v(e_{\text{Design}})$) are given for each sampling design in 3rd

Table 3.1 V_i and $\widehat{\sum_{i=1}^{N} V_i}$, for Different Qualitative and Quantitative RR Models

RR	V_i	$\widehat{\sum_{i=1}^{N} V_i}$
Warner	$\frac{p(1-p)}{(2p-1)^2}$ (Known)	$\frac{Np(1-p)}{(2p-1)^2}$
URL	$\frac{(1-p_1)(1-p_2)(p_1+p_2-2p_1p_2)}{(p_1-p_2)^2}(y_i - x_i)^2$	$e_{\text{Design}}(V_i)$ E.g. $\sum_{i=1}^{n} \frac{v_i}{\pi_i}$ (Design: HT estimator)
Kuk	$\frac{(1-\theta_1-\theta_2)}{k(\theta_1-\theta_2)} y_i + \frac{\theta_2(1-\theta_2)}{k(\theta_1-\theta_2)^2}$	$\frac{(1-\theta_1-\theta_2)}{k(\theta_1-\theta_2)} e_{\text{Design}} + \frac{N\theta_2(1-\theta_2)}{k(\theta_1-\theta_2)^2}$
Christofides	$\frac{\sum_{K=1}^{M} K^2 p_K - \mu^2}{(M+1-2\mu)^2}; \mu = \sum_{K=1}^{M} K p_K$ (known)	$\frac{N(\sum_{K=1}^{M} K^2 p_K - \mu^2)}{(M+1-2\mu)^2}$
Boruch	$\frac{(p_2-p_1)}{(1-p_1-p_2)} y_i + \frac{p_1(1-p_1)}{(1-p_1-p_2)^2}$	$\frac{(p_2-p_1)}{(1-p_1-p_2)} e_{\text{Design}} + \frac{Np_1(1-p_1)}{(1-p_1-p_2)^2}$
Mangat and Singh	$\frac{(1-T)(1-p)[T+(1-T)p]}{(T+(1-T)(2p-1))^2}$ (known)	$\frac{N(1-T)(1-p)[T+(1-T)p]}{(T+(1-T)(2p-1))^2}$
Device I	$\frac{\sigma^2}{\mu^2} y_i^2 + \frac{\psi^2}{\mu^2}$	$\frac{\sigma^2}{\mu^2} e_{\text{Design}}(y^2) + N \frac{\psi^2}{\mu^2}$
Device II	$\alpha y_i^2 + \beta y_i + \gamma$; where $\alpha = C(1-C)$, $\beta = -2C(\sum_{j=1}^{M} q_j x_j)$, $\gamma = \sum q_j x_j^2 - (\sum q_j x_j)^2$	$\alpha e_{\text{Design}}(y^2) + \beta e_{\text{Design}} + N\gamma$

Table 3.2 Numerical study for HH estimator

RR	Parameter values $(n = 30)$	e_{HH}	$\frac{v(e_{SRSWR,RR})}{1 + (\frac{N}{n})11}$	1	$\frac{11}{(\sum_{i=1}^{N} V_i)}$	$v(e_{HH})$	G_{RR}	Relative G_{RR}
Warner	$p = 0.3$	94.01	672.14	83.44	152.25	435.66	236.48	35.18
URL	$(p_1, p_2) = (0.25, 0.55)$	99.39	618.58	70.39	141.77	460.75	157.83	25.51
Kuk	$(\theta_1, \theta_2, k) =$ (0.4, 0.7, 2)	90.94	671.77	89.87	150.48	417.03	254.73	37.91
Christofides	$(p_1, p_2, p_3,$ $p_4, k) =$ (0.2, 0.4, 0.3, 0.1, 4)	112.34	2317.74	47.03	587.25	999.72	1318.01	56.86
Boruch	$(p_1, p_2) = (0.3, 0.4)$	86.08	1261.52	103.99	299.36	544.70	716.82	56.82
Mangat and Singh	$(T, p) = (0.3, 0.45)$	110.17	2059.6	52.01	519.21	918.13	1141.46	55.42
Device I	$(\mu, \gamma) = (4.75, 25)$	98,030.44	402,578.858	213,281.007	48,956.341	359,017.542	43,561.316	10.82
Device II	$C = 0.4$	97,313.65	1,286,093.324	931,194.002	91,784.307	401,122.565	884,970,759	68.81

Table 3.3 Numerical study for Des Raj estimator

RR	Parameter value ($n = 30$)	e_D	$\dfrac{v(N\bar{r})}{1 + (\frac{N}{n})II}$	I	$\dfrac{II}{(\sum_{i=1}^{N} V_i)}$	$v(e_D)$	G_{RR}	Relative G_{RR}
Warner	$p = 0.3$	86.07	667.08	78.38	152.25	567.96	99.11	14.86
URL	$(p_1, p_2) = (0.25, 0.55)$	83.45	591.32	79.63	132.33	478.24	113.08	19.12
Kuk	$(\theta_1, \theta_2, k) =$ $(0.4, 0.7, 2)$	84.83	657.82	79.87	149.47	559.66	98.16	14.92
Christofides	$(p_1, p_2, p_3,$ $p_4, k) =$ $(0.2, 0.4, 0.3, 0.1, 4)$	115.36	2325.04	54.34	587.25	2107.05	217.99	9.38
Boruch	$(p_1, p_2) = (0.30, 0.40)$	107.72	1233.95	48.53	306.57	1055.10	178.85	14.49
Mangat and Singh	$(T, p) = (0.3, 0.45)$	101.05	2089.27	81.68	519.20	1765.91	323.35	15.47
Device I	$(\mu, \gamma) = (5.75, 22.5)$	107,922.5	285,618,819	143,827,812	36,670,088	239,819,437	45,799,382	16.03
Device II	$C = 0.4$	105,046.6	451,243,549	274,215,550	45,783,103	422,159,195	29,084,354	6.44

Table 3.4 Numerical illustration for RHC estimation

RR	Parameter values ($n = 35$)	e_{RHC}	$v_{SRSWR, RHC}$ $1 + \left(\frac{N}{n}\right)11$	1	11 $\widehat{\sum_{i=1}^{N} V_i}$	$v(e_{RHC})$	G_{RR}	Relative G_{RR}
Warner	$p = 0.4$	89.46	2402.63	95.86	696	2389.42	13.18	0.54
URL	$(p_1, p_2) = (0.25, 0.75)$	91.15	181.07	49.06	39.83	173.51	7.55	4.17
Kuk	$(\theta_1, \theta_2, k) = (0.6, 0.2, 2)$	81.23	322.01	62.47	78.30	280.21	41.81	12.98
Christofides	$(p_1, p_2, p_3,$ $p_4, k) = (0.3, 0.4, 0.2, 0.1, 4)$	110.77	557.01	22.37	161.31	532.73	24.28	4.36
Boruch	$(p_1, p_2) = (0.2, 0.5)$	85.26	295.38	58.57	71.45	289.62	5.76	1.95
Mangat and Singh	$(T, p) = (0.6, 0.8)$	100.01	73.71	33.61	12.09	70.88	2.83	3.84
Device I	$(\mu, \gamma) = (5.75, 21.75)$	113,583.7	267,462,795	131,684,044	40,967,727	254,475,090	12,987,705	4.85
Device II	$C = 0.5$	106,056	476,137,721	292,847,416	55,303,110	472,565,929	3,571,793	0.75

Table 3.5 Numerical illustration for LMS, HT estimator

RR	Parameter value ($n = 35$)	e_{LMS}	$v_{SRSWOR,LMS}$ $1 + \left(\frac{N}{n}\right) II$	I	II $\widehat{\sum_{i=1}^{N} V_i}$	$v(e_{LMS})$	G_{RR}	Relative G_{RR}
Warner	$p = 0.4$	99.31	2388.15	81.41	696	2387.89	0.25	0.01
URL	$(p_1, p_2) = (0.25, 0.65)$	93.08	409.80	49.37	108.74	320.26	89.53	21.84
Kuk	$(\theta_1, \theta_2, k) = (0.6, 0.2, 2)$	103.55	310.02	31.99	83.88	300.40	9.61	3.10
Christofides	$(p_1, p_2, p_3,$ $p_4, k) = (0.3, 0.3, 0.3, 0.1, 4)$	93.87	1087.19	61.97	309.33	1002.25	84.93	7.81
Boruch	$(p_1, p_2) = (0.2, 0.2)$	110.45	1083.69	34.12	316.68	1083.42	0.27	0.02
Mangat and Singh	$(T, p) = (0.6, 0.3)$	99.42	442.41	42.07	120.79	442.13	0.28	0.06
Device I	$(\mu, \gamma) = (5.75, 21.75)$	113,985.4	233,150,263	105,705,536	38,453,150	232,870,526	279,736.9	0.12
Device II	$C = 0.5$	118,392.7	612,103,074	390,826,077	66,764,611	611,847,676	255,398.2	0.04

Table 3.6 Numerical illustration for LMS-ratio estimator

RR	Parameter value ($n = 40$)	e^*_{LMS}	$\dfrac{v_{SRSWOR,LMS*}}{1+\left(\frac{N}{n}\right)II}$	I	$\dfrac{II}{\widehat{\left(\sum_{i=1}^N V_i\right)}}$	$v\left(e^*_{LMS}\right)$	G_{RR}	Relative G_{RR}
Warner	$p = 0.7$	103.14	470.98	29.46	152.25	457.35	13.63	2.89
URL	$(p_1, p_2) = (0.4, 0.7)$	95.62	336.38	37.49	103.06	320.99	15.39	4.57
Kuk	$(\theta_1, \theta_2, k) = (0.4, 0.8, 3)$	103.32	186.36	24.28	55.89	160.41	25.95	13.92
Christofides	$(p_1, p_2, p_3, p_4, k) =$ $(0.23, 0.23, 0.23, 0.11, 0.2, 5)$	105.11	5357.82	94.82	1814.83	4594.71	763.10	14.24
Boruch	$(p_1, p_2) = (0.2, 0.3)$	93.19	309.48	40.13	92.88	303.92	5.56	1.79
Mangat and Singh	$(T, p) = (0.6, 0.4)$	85.23	271.36	44.44	78.25	67.04	204.31	75.29
Device I	$(\mu, \gamma) = (5.75, 22.5)$	101,523.4	221,444,964	115,614,057	12,555,103	210,169,224	11,275,740	5.09
Device II	$C = 0.6$	99,933.06	364,953,389	218,231,694	50,593,688	324,831,407	40,121,982	10.99

and 7th columns of the tables. Using the RR survey data at hand, unbiased estimate of variance for simpler alternative is computed and shown in the 4th column of Tables 3.2, 3.3, 3.4, 3.5 and 3.6. Here, "I" stands for the design-based estimated variance from PPS sample and "II" for the estimate of $\sum_{i=1}^{N} V_i$.

3.4 Conclusion

Based on the values of the estimated variances, it can be stated that PPSWR-HH estimator is significantly more gainful in efficiency than the SRSWR-Expansion estimator regardless of the RR models. Furthermore, PPSWOR-Des Raj estimator outperforms the SRSWOR-Expansion estimator. In comparison to Lahiri-Midzuno-Sen (LMS) sampling design-Ratio estimator and Horvitz-Thompson estimator, with SRSWOR-Expansion estimator, LMS-Ratio estimator is more beneficial for all RR model.

References

Adhikary, A. K. (2016). Variance estimation in randomizsed response surveys. In *Handbook of Statistics* (Vol. 34, pp. 191–207). Elsevier.

Boruch, R. F. (1972). Relations among statistical methods for assuring confidentiality of social research data. *Social Science Research, 1*(4), 403–414.

Chaudhuri, A. (2011). *Randomized response and indirect questioning techniques in surveys.* CRC Press.

Chaudhuri, A., & Christofides, T. C. (2013). *Indirect questioning in sample surveys.* Springer.

Chaudhuri, A., & Pal, S. (2022). *A comprehensive textbook on sample surveys.* Springer Nature.

Chaudhuri, A., & Patra, D. (2023). How to use randomized response survey data at hand by a specific procedure to judge its efficiency versus a possible rival. *Communications in Statistics-Theory and Methods.* https://doi.org/10.1080/03610926.2023.2250489

Chaudhuri, A., & Samaddar, S. (2022). Estimating the population mean using a complex sampling design dependent on an auxiliary variable. *Statistics in Transition New Series, 23*(4), 39–54.

Chaudhuri, A., Christofides, T. C., & Saha, A. (2009). Protection of privacy in efficient application of randomized response techniques. *Statistical Methods & Applications, 18*, 389–418.

Chaudhuri, A., Christofides, T. C., & Rao, C. R. (2016). *Handbook of statistics, data gathering, analysis and protection of privacy through randomized response techniques: qualitative and quantitative human traits* (Vol. 34). NL: Elsevier.

Christofides, T. C. (2003). A generalized randomized response technique. *Metrika, 57*, 195–200.

Cochran, W. G. (1953). *Sampling techniques* (1st ed.). Wiley.

Cochran, W. G. (1963). *Sampling techniques* (2nd ed.). Wiley.

Cochran, W. G. (1977). *Sampling techniques* (3rd ed.). Wiley.

Des, R. (1956). Some estimators in sampling with varying probabilities without replacement. *Journal of American Statistical Association, 51*, 269–284.

Eriksson, S. C. (1973). A new model for RR. *International Statistical Review, 41*, 101–113.

Greenberg, B. G., Abul-Ela, A. L., Simmons, W. R., & Horvitz, D. G. (1969). The unrelated question randomized response model: Theoretical framework. *Journal of American Statistical Association, 64*, 520–539.

Hansen, M. M., & Hurwitz, W. N. (1943). On the theory of sampling from finite populations. *Annals of Mathematical Statistics, 14*, 333–362.

Horvitz, D. G., & Thompson, D. J. (1952). A generalization of sampling without replacement from a finite universe. *Journal of American Statistical Association, 47*, 663–685.

Horvitz, D. G., Shah, B. V., & Simmons, W. R. (1967). The unrelated question randomized response model. In *Proceedings of Social Statistics Section, American Statistical Association*, (pp. 65–72).

Kuk, A. Y. (1990). Asking sensitive questions indirectly. *Biometrika, 77*(2), 436–438.

Lahiri, D. B. (1951). A method of sample selection providing unbiased ratio estimates. *Bulletin of International Statistical Institute, 3*(2), 133–140.

Mangat, N. S., & Singh, R. (1990). An alternative randomized response procedure. *Biometrika, 77*, 439–442.

Midzuno, H. (1949). An outline of the theory of sampling systems. *Annals of the Institute of Statistical Mathematics, 1*, 149–156.

Rao, J. N. (1961). On the estimate of variance in unequal probability sampling. *Annals of the Institute of Statistical Mathematics, 13*, 57–60.

Rao, J. N., Hartley, H. O., & Cochran, W. C. (1962). On a simple procedure of unequal probability sampling without replacement. *Journal of Royal Statistical Society, Series B, 24*, 482–491.

Sen, A. R. (1953). On the estimator of the variance in sampling with varying probabilities. *Journal of Indian Society of Agricultural Statistics, 5*, 119–127.

Warner, S. L. (1965). Randomized response: A survey technique for eliminating evasive answer bias. *Journal of American Statistical Association, 60*, 63–69.

Chapter 4
Fixing the Size of a Varying Probability Sample in a Direct and a Randomized Response Survey

4.1 Introduction

In this chapter, our intention is to estimate the proportion of population in a community bearing a stigmatizing characteristic A. Suppose y is a stigmatizing variable defined on a finite population $U = (1, 2, ..., i, ...N)$. This real variable y takes the values y_i which may be either 1 or 0 respectively for a person i in U bearing the sensitive feature A or its complement A^c. Let us unbiasedly estimate the population total $Y = \sum_{i=1}^{N} y_i$ or mean $\overline{Y} = \frac{Y}{N}$ with the help of a sample s of a 'suitable size n' from U. The sample s is to be chosen in accordance with a sampling design p assigning a value $p(s)$ to s. Direct response surveys, also known as DR surveys or randomized response (RR) technique, also known as RRT, are to be executed on the chosen units. Simple Random Sampling With Replacement (SRSWR) is the simplest design.

Here, we shall discuss more complex sampling designs including Proportional to Size With Replacement (PPSWR), Inclusion Probability Proportional to Size (IPPS) and Rao-Hartley-Cochran (RHC) sampling scheme in addition to the simplest design with its variant Simple Random Sampling Without Replacement (SRSWOR). In Sect. 4.2, we decide to address a few RRTs. In Sect. 4.3, estimation procedures provided by Hansen-Hurwitz (HH), Horvitz-Thompson (HT) and by RHC themselves will be discussed. Our main focus is on how to specify sample-size for the respective sampling designs to be followed in DR and RR surveys. Chaudhuri and Patra (2023a, 2023b) in their articles discussed the strategy to prescribe sample-size, covered in this chapter.

4.2   A Few Illustrative RR Devices

4.2.1   *Warner's RR Device*

Warner (1965), the pioneer of RRTs, suggested an interviewer to gather truthful response from a sampled person i of U through an RR device as

$I_i = 1$ if there is a 'match' with the sampled person $i's$ feature A or A^c when he/she draws a card randomly from a deck of cards, containing a large number of identical cards marked A or A^c in proportions $p : (1 - p)$, $\left(0 < p < 1, p \neq \frac{1}{2} \right)$.

$= 0$ if there is 'no' match.

Considering E_R and V_R generically as expectation and variance operators, we get

$$E_R(I_i) = py_i + (1 - p)(1 - y_i), i \in U \text{ and}$$
$$V_R(I_i) = E_R\left(I_i^2\right) - E_R^2(I_i) = E_R(I_i)(1 - E_R(I_i))$$
$$= p(1 - p), \text{ since } I_i^2 = I_i \text{ and } y_i^2 = y_i.$$

Then, $E_R(r_i) = y_i$ and $V_R(r_i) = \frac{p(1-p)}{(2p-1)^2}$ $\forall i$ in U if $r_i = \frac{I_i - (1-p)}{(2p-1)}$.

4.2.2   *URL RRT*

In this RRT, an RR emerges from a sampled person i of U as

$I_i = 1$ if a 'match' results in i's true nature namely, the stigmatizing A or an unrelated innocuous feature B when he/she on request randomly draws a card from a deck of cards marked A or B in proportions $p_1 : (1 - p_1)$, $(0 < p_1 < 1)$.

$= 0$ if 'no match'.

Another RR from this respondent emerges independently as

$J_i = 1$ if there is a 'match' when the respondent on request draws similarly a card from a second box, containing a large number of cards A and B in proportions $p_2 : (1 - p_2), 0 < p_2 < 1$ but $p_1 \neq p_2$.

$= 0$ if there is 'no match'.

Then, taking $r_i = \frac{p_2 I_i - p_1 J_i}{p_1 - p_2}$, one may get

$$E_R(r_i) = y_i \text{ and } V_R(r_i)$$
$$= \frac{(1 - p_1)(1 - p_2)(p_1 + p_2 - 2p_1 p_2)}{(p_1 - p_2)^2}(y_i - x_i)^2 = V_i$$

with

$$x_i = 1 \text{ if } i \text{ bears } B$$
$$= 0 \quad \text{if } i \text{ bears } B^c, \text{ the complement of } B.$$

4.2.3 Kuk's RRT

Let f_i be the number of red cards on choosing $k(> 1)$ cards from either box by SRSWR. The cards are chosen from the box with red and non-red cards in proportions $\theta_1 : (1 - \theta_1), 0 < \theta_1 < 1$ if i bears A. Cards are chosen from another similar box with the red: non-red in proportions $\theta_2 : (1 - \theta_2), \theta_1 \neq \theta_2$, if i bears A^c. Here, the interviewer derives the RR as f_i with

$$E_R(f_i) = k[y_i\theta_1 + (1 - y_i)\theta_2] = k[\theta_2 + y_i(\theta_1 - \theta_2)] \text{ and}$$
$$V_R(f_i) = k[y_i\theta_1(1 - \theta_1) + (1 - y_i)\theta_2(1 - \theta_2)]$$
$$= k[\theta_2(1 - \theta_2) + y_i(\theta_1 - \theta_2)].$$

Then, taking $r_i(k) = \frac{\frac{f_i}{k} - \theta_2}{\theta_1 - \theta_2}$ we get

$$E_R(r_i(k)) = y_i \text{ and } V_R(r_i(k)) = V_i(k), \text{ say}$$
$$= b_i(k)y_i + c_i(k),$$

where $b_i(k) = \frac{1 - \theta_1 - \theta_2}{k^2(\theta_1 - \theta_2)^2}$ and $c_i(k) = \frac{\theta_2(1 - \theta_2)}{k^2(\theta_1 - \theta_2)^2}$.

4.2.4 Forced Response RRT

A sampled person i from U is approached with a box containing large number of cards marked 'yes', 'no' and 'genuine' in respective proportions p_1, p_2 and $(1 - p_1 - p_2), 0 < p_1, p_2 < 1, p_1 + p_2 < 1, p_1 \neq p_2$. On request, the sampled person responds.

$I_i = 1$ if sampled person i randomly draws a card marked 'genuine' and his/her feature is A.

or the person randomly chooses, a card marked 'yes'.

$= 0$ if i draws a card marked 'no' or he/she draws a card marked 'genuine' and bears A^c.

Then, taking $r_i = \frac{l_i - p_1}{1 - p_1 - p_2}$, we get $E_R(r_i) = y_i$ and $V_R(r_i) = \frac{p_1(1-p_1) + y_i(1-p_1-p_2)(p_2-p_1)}{(1-p_1-p_2)^2} = V_i$.

4.2.5 Device I

Let us take Chaudhuri's (2011) RR device in the context of quantitative stigmatizing feature like duration of a stay behind the bar for a criminal offence as follows.

Let a sampled respondent i in U be approached with two boxes. The 1st box contains similar cards bearing numbers $a_1, a_2, \ldots a_j, \ldots, a_T$ with mean $\mu_a = \frac{1}{T}\sum_{j=1}^{T} a_j \neq 0$ and variance $\sigma_a^2 = \frac{1}{T-1}\sum_{j=1}^{T}(a_i - \mu_a)^2$. The cards in other box bear numbers $b_1, \ldots, b_k, \ldots, b_M$ having mean $\mu_b = \frac{1}{M}\sum_{k=1}^{M} b_k$ and variance $\sigma_b^2 = \frac{1}{M-1}\sum_{k=1}^{M}(b_k - \mu_b)^2$.

On request, the sampled person i is to randomly draw a card from box 1 and find the number a_j, written on the card. He/she also independently draws a card at random from the other box and finds, say, as labelled b_k.

Then, if the sampled person bears the value y_i, say, of the variable of interest as say y, the RR from i is to be recorded as $z_i = a_j y_i + b_k$.

Then, $E_R(z_i) = \mu_a y_i + \mu_b$ and $V_R(z_i) = \sigma_a^2 y_i^2 + \sigma_b^2$.

Then, taking $r_i = \frac{z_i - \mu_b}{\mu_a}$, one may get $E_R(r_i) = y_i$ and $V_R(r_i) = y_i^2 \frac{\sigma_a^2}{\mu_a^2} + \frac{\sigma_b^2}{\mu_a^2} = V_i$, say.

4.2.6 Eriksson's RRT

A sampled person i from U is approached with a proportion, say, $C(0 < C < 1)$ of cards marked 'correct' and the remaining cards bear a real number $z_1, z_2, \ldots, z_m$ with known proportions $q_1, q_2, \ldots, q_m$, respectively, such that $\sum_{j=1}^{m} q_j = 1 - C(0 < q_j < 1 \forall j)$.

$$S_i = y_i \text{ with probability } C.$$
$$= z_j \text{ with probability } q_j.$$

Then, $r_i = \frac{S_i - \sum_{j=1}^{m} q_j z_j}{C}$ has $E_R(r_i) = y_i$ and $V_R(r_i) = \frac{1}{C^2} V_R(S_i) = ay_i^2 + by_i + L$, where a, b, L are known constants.

4.3 A Few Illustrative Varying Probability Sampling Schemes

4.3.1 Probability Proportional to Size with Replacement (PPSWR) Sampling

Let x denote a size-measure variable with values $x_i (> 0 \, \forall i \in U)$ of ith unit, supposed to be well and positively correlated with the variable of interest y. In addition, $X = \sum_{i=1}^{N} x_i$ and $p_i = \frac{x_i}{X}$, the normed size-measures of the units i are known to the investigator.

Now, the units i are selected with probabilities $p_i, i = 1, 2 \ldots N$ in n independent draws from U.

Then, for a direct survey, the Hansen-Hurwitz (1943) unbiased estimator for $Y = \sum_{i=1}^{N} y_i$ is

$$t_{HH} = \frac{1}{n} \sum_{k=1}^{n} \frac{y_k}{p_k},$$

denoting y_k, p_k as the values of y_i, p_i for the unit chosen on the kth draw, $k = 1, 2 \ldots n$.

Then, $V(t_{HH}) = \frac{1}{n} \left(\sum_{i=1}^{N} \frac{y_i^2}{p_i} - Y^2 \right)$ and

$$v(t_{HH}) = \frac{1}{2n^2(n-1)} \sum_{k \neq}^{n} \sum_{k'}^{n} (\frac{y_k}{p_k} - \frac{y_{k'}}{p_{k'}})^2.$$

It is obvious that $E_p v(t_{HH}) = V(t_{HH})$.

Now, denoting E_p and V_p generically as the sampling design-based expectation and variance operators, one may get $E = E_P E_R = E_R E_P$ and $= E_P V_R + V_P E_R = E_R V_P + V_R E_P$, the overall expectation and variance operators.

Using RR survey data corresponding to t_{HH} an unbiased estimator for Y is

$$e_{HH} = \frac{1}{n} \sum_{k=1}^{n} \frac{r_k}{p_k},$$

where r_k denotes the value of r_i for the unit i chosen on the kth draw.

Then, for Warner's RRT

$$V(e_{HH}) = V(t_{HH}) + \frac{p(1-p)}{n(2p-1)^2} \sum_{i=1}^{N} \frac{1}{p_i}$$

writing V_k for V_i for the unit i chosen on the kth draw.

4.3.2   Inclusion Probability Proportional to Size (IPPS) Sampling

The IPPS sampling schemes described by Brewer and Hanif (1983) and Chaudhuri and Vos (1988) are numerous. Let z_i be certain known positive numbers. Suppose, on the 1st draw, a unit i of U is selected with a probability proportional to z_i, and on the 2nd draw, a unit $j(\neq i)$ of U is selected with a probability proportional to z_j and out of the remaining $(N-2)$ units, by SRSWOR $(n-2)$ units are chosen.

Then, the selection probability of such a sample s of size n is

$$p(s) = \frac{z_i}{Z} \frac{z_j}{Z - z_i} \frac{1}{\binom{N-2}{n-2}}, \text{ where } Z = \sum_{i=1}^{N} z_i.$$

In addition, the inclusion probability of i in such a sampling scheme is given below.

For the 1st two draws, inclusion probability of i is

$\pi_i(2) = \frac{z_i}{Z} + \sum_{i \neq j}^{N} \frac{z_j}{Z - z_j} \frac{z_i}{Z}$, and hence, by this scheme,

$\pi_i(n) = \pi_i(2) + (1 - \pi_i(2))\frac{n-2}{N-2}$ is the inclusion probability in the entire sample of size n.

Taking $Q_i = \frac{z_i}{Z}$, we may get $\pi_{ij}(2) = \frac{Q_i Q_j}{1 - Q_i} + \frac{Q_i Q_j}{1 - Q_j}$ as the inclusion probability of i and j in the 1st two draws, in this scheme.

In this scheme, now, in the entire sample of size, the inclusion probability of i and j both is

$$\pi_{ij}(n) = \pi_{ij}(2) + \left(\frac{n-2}{N-2}\right)\left(\pi_i(2) + \pi_j(2) - 2\pi_{ij}(2)\right)$$
$$+ \left(\frac{n-2}{N-2}\right)\left(\frac{n-3}{N-3}\right)\left(1 - \pi_i(2) - \pi_j(2) + \pi_{ij}(2)\right).$$

If $\pi_i(n)$ here is equated to np_i, then the scheme is called IPPS sampling scheme.

For any sampling scheme, an unbiased estimator for Y is the Horvitz-Thompson (HT) estimator

$$t_{HT} = \sum_{i \in s} \frac{y_i}{\pi_i}$$

where the inclusion probability of i is $\pi_i(> 0)$ and $\sum_{i=1}^{N} \pi_i = n$.

Then,

$$E_P(t_{HT}) = Y$$

$$V_P(t_{HT}) = \sum_{i<}^{N} \sum_{j=1}^{N} (\pi_i \pi_j - \pi_{ij}) \left(\frac{y_i}{\pi_i} - \frac{y_j}{\pi_j} \right)^2 \text{ and}$$

$$v_p(t_{HT}) = \sum_{i<} \sum_{j\in s} \frac{(\pi_i \pi_j - \pi_{ij})}{\pi_{ij}} \left(\frac{y_i}{\pi_i} - \frac{y_j}{\pi_j} \right)^2$$

assuming every sample s has only distinct units and the number of units in s is a fixed number.

For an RR survey, corresponding HT estimator is $e_{HT} = \sum_{i\in s} \frac{r_i}{\pi_i}$, with $E_R(r_i) = y_i$. The estimator e_{HT} is unbiased for Y in the sense $E(e_{HT}) = E_P(t_{HT}) = E_R\left(\sum_{i=1}^{N} r_i \right) = Y$ and the variance of e_{HT} is $V(e_{HT}) = V_P(t_{HT}) + \sum_{i=1}^{N} \frac{V_i}{\pi_i}$.

Then,

$$v(e_{HT}) = \sum_{i<} \sum_{j\in s} \frac{(\pi_i \pi_j - \pi_{ij})}{\pi_{ij}} \left(\frac{r_i}{\pi_i} - \frac{r_j}{\pi_j} \right)^2 + \sum_{i\in s} \frac{v_i}{\pi_i}$$

4.3.3 Rao, Hartley and Cochran (RHC) Sampling

Here the population $U = (1, 2, \ldots i \ldots N)$ is randomly split up into n disjoint parts. An SRSWOR of N_1 units is chosen to form the 1st group and then successively taking $(n - 1)$ more SRSWORs mutually exclusively of sizes $N_2, N_3, \ldots, N_n$ such that $\sum_n N_i = N$, $\sum_n$ denoting sum over the n disjoint groups thus formed. Then, p_{ij} values of p_i's for the respective groups $i = 1, 2 \ldots n$ are noted, and from each of the n groups, one unit j of the N_i units is chosen with the probability $\frac{p_{ij}}{Q_i}$, where $Q_i = \sum_{j=1}^{N_i} p_{ij}$, and this is independently repeated for all the n groups. Then,

$$t_{RHC} = \sum_n y_i \frac{Q_i}{p_{ij}}$$

is taken as an unbiased estimator for $Y = \sum_n Y_i$ where $Y_i = \sum_{j=1}^{N_i} y_{ij}$ and y_i is the value of the unit selected from the ith group with the units $y_{i1}, y_{i2}, \ldots, y_{iN_i}$.

Then, it follows that

$$V(t_{\mathrm{RHC}}) = \frac{\sum_n N_i^2 - N}{N(N-1)} \sum_n \sum_n p_i p_j \left(\frac{y_i}{p_i} - \frac{y_j}{p_j}\right)^2$$

and

$$v(t_{\mathrm{RHC}}) = \frac{\sum_n N_i^2 - N}{N^2 - \sum_n N_i^2} \sum_n \sum_n Q_i Q_j \left(\frac{y_i}{p_i} - \frac{y_j}{p_j}\right)^2$$

is an unbiased estimator of $V(t_{\mathrm{RHC}})$.

A suitable choice of the N_i's is $N_i = \left[\frac{N}{n}\right]$, where $\left[\frac{N}{n}\right]$ is the integer part of N divided by n for $i = 1, \ldots m$ and $N_i = \left[\frac{N}{n}\right] + 1$ for $i = m+1, \ldots, n$ such that $\sum_{i=1}^{m} N_i + \sum_{i=m+1}^{n} N_i = N$.

For RR survey data, r_i based on RHC, an unbiased estimator for Y is

$$e_{\mathrm{RHC}} = \sum_n r_i \frac{Q_i}{p_{ij}} \quad \text{and}$$

$$v(e_{\mathrm{RHC}}) = v(t_{\mathrm{RHC}}) + \sum_n v_i \frac{Q_i}{p_{ij}}$$

where $E_R(r_i) = y_i$ and v_i is the unbiased estimator of $V_R(r_i)$.

4.4 Sample-Sizes: DR, RR Surveys

The following have been suggested by Chaudhuri (2010, 2014, 2018, 2020), Chaudhuri and Dutta (2018), and Chaudhuri and Sen (2020) in terms of sample-size specification.

Suppose t is an unbiased estimator for a finite population total Y and our intention is to choose t as so accurate that

$$\mathrm{Prob}[|t - Y| < fY] \geq 1 - \alpha,$$

choosing f as proper fraction like $0.1, 0.2$, etc. and α is a positive quantity so small as, say, $0.05, 0.01$, etc.

Chebyshev's inequality says

$$\mathrm{Prob}\left[|t - Y| < \lambda\sqrt{V(t)}\right] \geq 1 - \frac{1}{\lambda^2}$$

where λ is a positive number greater than 1.

Combining the above two inequalities, one may take

$$fY = \lambda\sqrt{V(t)} \text{ and } \alpha = \frac{1}{\lambda^2}.$$

giving us

$$100f = \frac{1}{\sqrt{\alpha}}\text{CV}(t) \qquad (4.1)$$

writing $\text{CV}(t) = 100\frac{\sqrt{V(t)}}{Y}$.

This $\text{CV}(t)$ is the coefficient of variation of t.

Below, we show how this (4.1) might aid us in recommending an appropriate size n of a sample to select in a certain sample-selection situation.

4.4.1 Sample-Size in Direct Surveys with SRSWR and SRSWOR

For SRSWR with $N\bar{y}$ to estimate Y or $\bar{y}$ to estimate $\bar{Y} = \frac{Y}{N}$,

$$V(\bar{y}) = \frac{\sigma^2}{n} = \frac{N-1}{Nn}S^2$$

writing $\sigma^2 = \frac{1}{N}\sum_{i=1}^{N}(y_i - \bar{Y})^2 = \frac{N-1}{N}S^2$ implying $S^2 = \frac{1}{N-1}\sum_{i=1}^{N}(y_i - \bar{Y})^2$.

The coefficient of variation of $\bar{y}$ is $\text{CV}(\bar{y}) = 100\sqrt{\frac{N-1}{Nn}}\frac{S}{\bar{Y}}$.

Writing $\text{CV} = 100\frac{S}{\bar{Y}}$, the coefficient of variation of all the N values of y_i's in the population, a rule to choose the sample-size is using (4.1)

$$n = \frac{(N-1)(\text{CV})^2}{N\alpha f^2}. \qquad (4.2)$$

For SRSWOR of size n, if $\bar{Y}$ is estimated by the sample mean $\bar{y}$, then $V(\bar{y}) = \frac{N-n}{Nn}S^2$ and therefore, using (4.1), an appropriate 'sample-size fixing rule' suggests

$$n = \frac{N}{1 + N\alpha f^2\left(\frac{100}{\text{CV}}\right)^2} \qquad (4.3)$$

as both (4.2) and (4.3) may be checked from Chaudhuri and Dutta (2018) and also Chaudhuri (2020).

Choosing N, f, α, CV and following (4.2), the values of n for an SRSWR may be worked out 'rounding it up' to the nearest positive integer. Similarly, for SRSWOR, specifying N, f, α and CV, the appropriate sample-size n may be chosen using (4.2) above rounding it up to the least positive integer.

Table 4.1 Sample-sizes for SRSWR and SRSWOR

N	CV	n by (4.2)	n by (4.3)
60	0.08	13	11
80	0.1	20	16

Thus, we construct Table 4.1, considering $\alpha = 0.05$ and $f = 0.1$.

The ratios $\frac{n}{N}$ look reasonable for both SRSWR and SRSWOR. These results are relevant to direct response (DR) surveys. Now let us turn to RR surveys.

4.4.2   Sample-Size in RR Surveys with SRSWR and SRSWOR

Warner's RRT

If an SRSWR of size n is taken from U, then $\bar{r}$, the sample mean of the r_i's has $E_R(\bar{r}) = \bar{y}$, the sample mean of the y_i's and $V_R(\bar{r}) = \frac{p(1-p)}{n(2p-1)^2}$.

Considering E_P and V_P as design-based expectation, variance operators and the overall expectation, variance operators as $E = E_P E_R = E_R E_P$ and $V = E_P V_R + V_P E_R = E_R V_P + V_R E_P$, one may get

$$E(\bar{r}) = \bar{Y} = \theta, \text{ say, and}$$

$$V(\bar{r}) = \frac{1}{nN} \sum_{i=1}^{N} (y_i - \bar{Y})^2 + \frac{p(1-p)}{n(2p-1)^2}$$

$$= \frac{1}{n}\left[\frac{p(1-p)}{(2p-1)^2} + \frac{\theta(1-\theta)}{N} \right]. \tag{4.4}$$

Similarly, for an SRSWOR in n draws, $\bar{r}$ has

$$E(\bar{r}) = \bar{Y} = \theta, \text{ say and}$$

$$V(\bar{r}) = \frac{N-n}{nN(N-1)} \sum_{i=1}^{N} (y_i - \bar{Y})^2 + \frac{p(1-p)}{n(2p-1)^2}$$

$$= \frac{1}{n}\left[\frac{p(1-p)}{(2p-1)^2} + \frac{N}{N-1}\theta(1-\theta) \right] - \frac{\theta(1-\theta)}{N-1} \tag{4.5}$$

on noting $y_i^2 = y_i$ because $y_i = 1$ or 0 and $\sum_{i=1}^{N} y_i^2 - N\bar{Y}^2 = N\theta(1-\theta)$.

Therefore, Equation (4.1) gives in these two cases,

$$100f = \frac{1}{\sqrt{\alpha}} \frac{\sqrt{V(\bar{r})}}{\bar{Y}} = \frac{1}{\sqrt{\alpha}} \frac{\sqrt{V(\bar{r})}}{\theta}.$$

Then, Eqs. (4.4) and (4.5) give

$$100f = \frac{1}{\sqrt{\alpha}}\frac{1}{\theta}\left[\frac{1}{n}\left\{\frac{p(1-p)}{(2p-1)^2} + \frac{\theta(1-\theta)}{N}\right\}\right]^{1/2} \text{ for SRSWR}$$

and

$$100f = \frac{1}{\sqrt{\alpha}}\frac{1}{\theta}\left[\frac{1}{n}\left\{\frac{p(1-p)}{(2p-1)^2} + \frac{N}{N-1}\theta(1-\theta)\right\} - \frac{\theta(1-\theta)}{N-1}\right]^{1/2} \text{ for SRSWOR}$$

(4.6)

Thus, for SRSWR, the rule is

$$n = \frac{\left[\frac{p(1-p)}{(2p-1)^2} + \frac{\theta(1-\theta)}{N}\right]}{(100f)^2\theta^2\alpha},$$

(4.7)

and for SRSWOR, the rule is

$$n = \frac{\left[\frac{p(1-p)}{(2p-1)^2} + \frac{N}{N-1}\theta(1-\theta)\right]}{(100f)^2\theta^2\alpha + \frac{\theta(1-\theta)}{N-1}}$$

(4.8)

Though $\theta = \bar{Y}$ is the estimand parameter and can never be known, to get an insight into the possibilities of making a rational choice of n, one may construct a table choosing N, f, α, θ and p as Table 4.2. Here also, we have taken $\alpha = 0.05$ and $f = 0.1$, similar to the previous table.

Table 4.2 shows the results for n are awful and we **do not recommend this method for RR survey**.

Device I

For an SRSWR of size n, we have $E(\bar{r}) = \bar{Y}$ and $V(\bar{r}) = \frac{1}{n}\left[\frac{\sum_{i=1}^{N}V_i}{N} + \frac{(N-1)S^2}{N}\right]$ for the sample mean $\bar{r}$.

Clearly, $CV(\bar{r}) = \frac{100}{\sqrt{n}}\frac{\left[\frac{\sum_{i=1}^{N}V_i}{N} + \frac{(N-1)S^2}{N}\right]^{1/2}}{\bar{Y}}$.

So, Eq. (4.1) gives $100f = \frac{CV(\bar{r})}{\sqrt{\alpha}}$.

Table 4.2 Choosing n in SRSWR and SRSWOR in an RR survey by Warner's RRT

N	θ	p	n by (4.7)	n by (4.8)
60	0.2	0.55	124	123
80	0.2	0.55	124	123

So, $(100f)^2 = \dfrac{100^2}{\alpha} \dfrac{\left[\frac{\sum_{i=1}^{N} V_i}{N} + \frac{(N-1)S^2}{N} \right]}{n\bar{Y}^2}$.

Therefore, the rule for n is

$$n = \frac{\left[\left(\frac{100}{\bar{Y}}\right)^2 \frac{\sum_{i=1}^{N} V_i}{N} + \frac{N-1}{N}(CV)^2 \right]}{(100f)^2\alpha}. \tag{4.9}$$

Suppose the RR survey data are gathered by SRSWOR in n draws, then

$$V(\bar{r}) = \frac{N-n}{Nn}S^2 + \frac{1}{n}\frac{\sum_{i=1}^{N} V_i}{N}, \text{ and}$$

$$[CV(\bar{r})]^2 = \left(\frac{100}{\bar{Y}}\right)^2 \left[\frac{N-n}{Nn}S^2 + \frac{1}{n}\frac{\sum_{i=1}^{N} V_i}{N} \right]$$

Therefore, (1.3) gives the rule for n as

$$\begin{aligned}
(100f)^2\alpha &= [CV(\bar{r})]^2 \\
&= \left(\frac{100}{\bar{Y}}\right)^2 \left[\frac{N-n}{Nn}S^2 + \frac{1}{n}\frac{\sum_{i=1}^{N} V_i}{N} \right] \\
&= \frac{1}{n}\left[\left(\frac{100}{\bar{Y}}\right)^2 \frac{\sum_{i=1}^{N} V_i}{N} + \left(\frac{N-n}{N}\right)(CV)^2 \right] \\
&= \frac{1}{n}\left[\left(\frac{100}{\bar{Y}}\right)^2 \frac{\sum_{i=1}^{N} V_i}{N} + (CV)^2 \right] - \frac{(CV)^2}{N}
\end{aligned}$$

or

$$n = \frac{\left[\left(\frac{100}{\bar{Y}}\right)^2 \frac{\sum_{i=1}^{N} V_i}{N} + (CV)^2 \right]}{(100f)^2\alpha + \frac{(CV)^2}{N}}. \tag{4.10}$$

Therefore, to choose n for SRSWR and for SRSWOR, Table 4.3 is to be used. Here, we consider $\alpha = 0.05$ and $f = 0.1$.

Table 4.3 Choosing n in SRSWR (using (4.9)) and SRSWOR (using (4.10)) in RR survey by Chaudhuri (2011) device

N	$\bar{Y}$	CV	n by (4.9)	n by (4.10)
50	36	10	851	608
60	36	10	844	634
80	36	10	861	689
100	36	10	829	691

Looking at Tables 4.1, 4.2 and 4.3, it is clear that Table 4.1 is constructed very easily employing only some arbitrary but reasonable values of CV to arrive at a choice of n that naturally ends up being slightly larger for SRSWR than for SRSWOR. Additionally, in Tables 4.2 and 4.3, also n for SRSWR is greater than SRSWORs. But their construction is different because we need values of $\theta = \overline{Y}$, and in addition, we need CV for the quantitative case.

Although they are reasonable for DR, *the sample-sizes for RR appear as absurd for the present approach based on Chebyshev's procedure.* We could suggest the following remedy to get around this anomaly.

There are two terms: *I* for variance of RRs and *II* for variance of DR-based values, for the s each variance of the RR's estimator for the estimand parameter like the proportion, total or mean. As shown by Chaudhuri and Sen (2020), *I* far exceeds *II*, in magnitude. However, *I* has a little relation to the sample-selection procedure. Therefore, in order to control the magnitude of *II* by the procedure based on Chebyshev's ideas, we should choose the sample-size directly. Then, we should assess how reasonable the magnitude of *I* comes out in relation to the sampling design and the sample-size.

Now let us see how different RR devices fare keeping the design fixed only as SRSWR and SRSWOR.

Simmons's Unrelated Response Model or URL RRT

If an SRSWR in n draws is taken and RRs are observed, then $\overline{r} = \frac{1}{n} \sum_{i=1}^{n} r_i$ has $E_R(\overline{r}) = \overline{y} = \frac{1}{n} \sum_{i=1}^{n} y_i$, $E(\overline{r}) = \overline{Y} = \theta$ and $V(\overline{r}) = \frac{1}{n} \left[\frac{\theta(1-\theta)}{N} + \frac{\sum_{i=1}^{N} V_i}{N} \right]$ with V_i as in Sect. 4.2.2.

Now, using (4.1) one may derive n as in (4.9).

If an SRSWOR in n draws is taken and RR survey data are gathered, $\overline{r} = \frac{1}{n} \sum_{i=1}^{n} r_i$ may be used to unbiasedly estimate $\overline{Y} = \theta$ and derived $V(\overline{r}) = \frac{N-n}{Nn} S^2 + \frac{\sum_{i=1}^{N} V_i}{nN}$ with V_i as above.

Then, by (4.1), a rule for n is as in (4.10).

Thus,

$$n = \frac{\left(\frac{100}{\theta}\right)^2 \frac{\sum_{i=1}^{N} V_i}{N} + \frac{N-1}{N}(CV)^2}{(100f)^2 \alpha} \quad \text{for SRSWR and}$$

$$n = \frac{\left(\frac{100}{\theta}\right)^2 \frac{\sum_{i=1}^{N} V_i}{N} + (CV)^2}{(100f)^2 \alpha + \frac{(CV)^2}{N}} \text{for SRSWOR .}$$

Since V_i is difficult to anticipate in both the above formulae for n, V_i should be replaced by its unbiased estimator $v_i = r_i(r_i - 1)$ and $\sum_{i=1}^{N} V_i$ by $\frac{1}{n_0} \sum_{1}^{n_0} r_i(r_i - 1)$ in both the above formulae for n. Here, n_0 refers to an arbitrary size of an SRSWR and SRSWOR taken to get the RR data as above to find r_i and hence use $\frac{1}{n_0} \sum_{1}^{n_0} r_i(r_i - 1)$, $\sum_{1}^{n_0}$ denoting sum over the n_0 values of $r_i(r_i - 1)$. We use the same notation in the above two formulae for n for SRSWR and n for SRSWOR. We simply replace $\frac{\sum_{i=1}^{N} V_i}{N}$

by $\frac{1}{n_0} \sum_1^{n_0} r_i(r_i - 1)$ which is calculated respectively for an SRSWR of size n_0 and SRSWOR of size n_0, using the realized values of r_i for them.

Thus, finally,

$$n = \frac{\left(\frac{100}{\theta}\right)^2 \frac{\sum_1^{n_0} r_i(r_i-1)}{n_0} + \frac{N-1}{N}(CV)^2}{(100f)^2\alpha} \tag{4.11}$$

and

$$n = \frac{\left(\frac{100}{\theta}\right)^2 \frac{\sum_1^{n_0} r_i(r_i-1)}{n_0} + (CV)^2}{(100f)^2\alpha + \frac{(CV)^2}{N}}. \tag{4.12}$$

Then, we construct Table 4.4 taking $\alpha = 0.05$ and $f = 0.1$.

Kuk's RRT

Suppose an SRSWR in n draws produces RRs as $r_i(k)$, then $\bar{r}(k) = \frac{1}{n}\sum_{i=1}^n r_i(k)$ is unbiased for $\theta = \bar{Y} = \frac{1}{N}\sum_{i=1}^N y_i$ and have the variance $V(\bar{r}(k)) = \frac{1}{n}\left[\frac{\theta(1-\theta)}{N} + \frac{\sum_{i=1}^N V_i(k)}{N}\right]$.

If, instead, an SRSWOR in n draws is taken producing the RRs as $r_i(k)$'s, then $\bar{r}(k) = \frac{1}{n}\sum_{i=1}^n r_i(k)$ is an unbiased estimate of $\theta = \bar{Y}$ having variance as $V(\bar{r}(k)) = \frac{N-n}{Nn}S^2 + \frac{1}{nN}\sum_{i=1}^N V_i(k)$.

Then, using (4.3), for SRSWR appropriate sample-size is

$$n = \frac{\left(\frac{100}{\theta}\right)^2 \left(\frac{\sum_{i=1}^N V_i(k)}{N}\right) + \left(\frac{N-1}{N}\right)(CV)^2}{(100f)^2\alpha} \tag{4.13}$$

and for SRSWOR it is

$$n = \frac{\left(\frac{100}{\theta}\right)^2 \left(\frac{\sum_{i=1}^N V_i(k)}{N}\right) + (CV)^2}{(100f)^2\alpha + \frac{(CV)^2}{N}}. \tag{4.14}$$

A usable table then is, taking $k = 3$, $\alpha = 0.05$ and $f = 0.1$ (Table 4.5).

In (4.13) and (4.14) also $\frac{\sum_{i=1}^N V_i(k)}{N}$ should be replaced by $\frac{1}{n_0}\sum_1^{n_0} V_i(k)$ just as in URL.

Table 4.4 Sample-size for URL by SRSWR and SRSWOR

N	θ	CV	p_1	p_2	n_0	n by (4.11)	n by (4.12)
60	0.2	10	0.44	0.52	8	969,019	675,015
80	0.2	10	0.44	0.52	10	630,019	504,016

Table 4.5 Sample-size for Kuk RRT by SRSWR and SRSWOR

N	θ	CV	p_1	p_2	n_0	n by (4.13)	n by (4.14)
60	0.2	10	0.44	0.52	8	643,569	505,483
80	0.2	10	0.44	0.52	10	665,519	521,127

Table 4.6 Sample-size for forced response by SRSWR and SRSWOR

N	θ	CV	p_1	p_2	n_0	n by (4.15)	n by (4.16)
60	0.2	10	0.44	0.52	8	7,850,020	6,121,890
80	0.2	10	0.44	0.52	10	7,850,020	6,880,016

Forced Response RRT

If an SRSWR is taken in n draws, then $\bar{r} = \frac{1}{n} \sum_{i=1}^{n} r_i$ unbiasedly estimates $\theta = \overline{Y}$.
Also variance $V(\bar{r}) = \frac{1}{n}[\frac{\theta(1-\theta)}{N} + \frac{1}{N} \sum_{i=1}^{N} V_i]$ with $V_i = V_R(r_i)$ as in Sect. 4.2.4.

If, instead, an SRSWOR is taken and RRs are derived as above, an unbiased
estimator for $\theta = \overline{Y} = \frac{1}{N} \sum_{i=1}^{N} y_i$ is $\bar{r} = \frac{1}{n} \sum_{i=1}^{n} r_i$. But its variance is $V(\bar{r}) = \frac{N-n}{Nn} S^2 + \frac{1}{Nn} \sum_{i=1}^{N} V_i = \frac{N-n}{n(N-1)}\theta(1-\theta) + \frac{1}{Nn} \sum_{i=1}^{N} V_i$.

Considering (4.1), we get

$$n = \frac{\left(\frac{100}{\theta}\right)^2 \left(\frac{\sum_{i=1}^{N} V_i}{N}\right) + \left(\frac{N-1}{N}\right)(CV)^2}{(100f)^2\alpha} \tag{4.15}$$

$$n = \frac{\left(\frac{100}{\theta}\right)^2 \left(\frac{\sum_{i=1}^{N} V_i}{N}\right) + (CV)^2}{(100f)^2\alpha + \frac{(CV)^2}{N}}, \tag{4.16}$$

as the appropriate n's for SRSWR and SRSWOR, respectively.

Similar to the case of URL, here also the term $\frac{\sum_{i=1}^{N} V_i}{N}$ may be estimated by
$\frac{1}{n_0} \sum_{1}^{n_0} v_i$ taking $v_i = p_1(1 - p_1) + r_i(1 - p_1 - p_2)(p_2 - p_1)$ as an unbiased estimator for V_i on taking an initial SRSWR and another SRSWOR of a sample-size n_0
and with RRs found as above in that.

Then, $\frac{\sum_{i=1}^{N} V_i}{N}$ in (4.15) and (4.16) may be replaced by $\frac{1}{n_0} \sum_{1}^{n_0} v_i$. We construct
here Table 4.6 taking $\alpha = 0.05$ and $f = 0.1$.

4.4.3 Sample-Size in DR Surveys with Varying Probability Sampling

In case an SRSWR or an SRSWOR is chosen and $N\bar{y}$ with $\bar{y}$ as the sample mean
in n draws, in either case to unbiasedly estimate $\overline{Y}$, then $V(N\bar{y})$ being equal to

$N^2 \frac{\sigma^2}{n} = \frac{N^2}{n} \left(\frac{N-1}{N} \right) S^2$ for SRSWR, writing $S^2 = \frac{1}{N-1} \sum_{i=1}^{N} (y_i - \overline{Y})^2$, $\overline{Y} = \frac{Y}{N}$, it is possible to use the formula (4.1) above to fix n vis-a-vis N, f and α on speculating magnitudes of $100 \frac{S}{Y}$, the coefficient of variations from the N population values of $y_i's$ ($i = 1, 2 \ldots N$).

But, it is difficult to use these facilities to choose n if varying probability samples are surveyed to estimate Y. To get around this, Chaudhuri and Dutta (2018) proposed postulating the following simple model shown below, which connects y variable with an auxiliary variable x that may be well and positively correlated with y.

Let us consider the model

$$y_i = \beta x_i + \epsilon_i, \quad i \in U \tag{4.17}$$

with β as an unknown constant, $\epsilon_i's$ are independently distributed random variables with $E_m(\epsilon_i) = 0 \, \forall i$ and $V_m(\epsilon_i) = \sigma^2 x_i^g$ with $\sigma (> 0)$, an unknown constant and g an unknown constant such that $0 \leq g \leq 2$.

Now, using (4.1), we need

$$\text{Prob}[|t_{\text{HH}} - Y| \leq fY] \geq 1 - \alpha = 1 - \frac{V(t_{\text{HH}})}{f^2 Y^2}.$$

This implies,

$$\alpha = \frac{V(t_{\text{HH}})}{f^2 Y^2}. \tag{4.18}$$

But from the above, we get no clue to fix the sample-size n. Chaudhuri and Dutta (2018), therefore, prescribe to consider

$$\alpha = \frac{E_m V(t_{\text{HH}})}{f^2 E_m(Y^2)} \tag{4.19}$$

instead of (4.18). Here, E_m is the expectation operator under the model (4.17).

Similarly for t_{HT} and t_{RHC}, Equation (4.19) above as an analogue to find a suitable clue for n.

Let us work out

$$E_m V(t_{\text{HH}}) = \frac{\sigma^2}{n} \left[X \sum_{i=1}^{N} x_i^{g-1} - \sum_{i=1}^{N} x_i^g \right]$$

$$E_m(Y^2) = \beta^2 X^2 + \sigma^2 \sum_{i=1}^{N} x_i^g.$$

Further restricting the model (4.17) to assume x has the following density

$$f(x) = e^{-x}, x > 0.$$

A random sample of x_i values can be easily drawn from this exponential density. In order to estimate Y in a direct survey, we can therefore compute the following table fixing n for PPSWR sampling. Similar to the previous tables, we restrict $\alpha = 0.05$ and $f = 0.1$ (Table 4.7).

Next we work out

$$E_m V(t_{\text{HT}}) = \frac{\sigma^2}{n} \left[X \sum_{i=1}^{N} x_i^{g-1} - n \sum_{i=1}^{N} x_i^g \right].$$

Similar to the HH estimator, we tabulate Table 4.8 to provide sample-size for estimating Y by HT estimator in a direct survey restricting $\alpha = 0.05$ and $f = 0.1$ in Equation (4.19).

Now, for t_{RHC}, we calculate

$$E_m V(t_{\text{RHC}}) = \sigma^2 \frac{\sum_n N_i^2 - N}{N(N-1)} \left[X \sum_{i=1}^{N} x_i^{g-1} - \sum_{i=1}^{N} x_i^g \right].$$

Due to this, we tabulate Table 4.9 to provide the sample-size for estimating Y by RHC strategy in a direct survey.

Tables 4.7, 4.8 and 4.9 conclude that our approach of fixing sample-sizes in DR surveys employing varying probabilities sampling schemes is rather successful. The sampling fractions $\frac{n}{N}$'s are coming out quite elegant.

Now, let us examine what might happen in RR surveys.

Table 4.7 Fixing **n** for PPSWR sampling in DR surveys

N	σ^2	β	g	n by (4.19)
60	2	15	1.5	19
80	2	15	1	18

Table 4.8 Fixing n for HT estimator in DR surveys

N	σ^2	β	g	n by (4.19)
60	2	15	1.5	12
80	2	15	1	14

Table 4.9 Fixing n for RHC strategy in DR surveys

N	σ^2	β	g	n by (4.19)
60	2	15	1.5	12
80	2	15	1	14

4.4.4 Sample-Size in RR Surveys with Varying Probability Sampling

Let us apply (4.19) to the following situations

(i)   PPSWR [Warner, URL, Kuk, Forced response, Eriksson's] with HH estimate
(ii) IPPS [Warner, URL, Kuk, Forced response, Eriksson's] with HT estimate
(iii) RHC [Warner, URL, Kuk, Forced response, Eriksson's] with RHC estimate
      (Tables 4.10, 4.11 and 4.12).

$$E_m V(e_{HH}|\text{Warner}) = E_m V(t_{HH}) + \frac{1}{n}\sum_{i=1}^{N}\frac{1}{p_i}\frac{p(1-p)}{(2p-1)^2}$$

$$= \frac{\sigma^2}{n}\left[X\sum_{i=1}^{N}x_i^{g-1} - \sum_{i=1}^{N}x_i^{g}\right] + \frac{1}{n}\sum_{i=1}^{N}\frac{1}{p_i}\frac{p(1-p)}{(2p-1)^2}$$

Table 4.10 Fixing sample-size in PPSWR sampling to estimate finite population total in RR surveys applying (4.19) with $\alpha = 0.05$, $f = 0.1$

Warner's RR	N	σ^2	β	g	p	n
	60	2	15	1.5	0.35	96
	80	2	15	1	0.35	66
URL RRT	N	σ^2	β	g	p_1, p_2	n
	60	2	15	1.5	0.35,0.65	2439
	80	2	15	1	0.35,0.65	2442
Kuk (taking $k = 3$)	N	σ^2	β	g	θ_1, θ_2	n
	60	2	5	1.5	0.35,0.65	21
	80	2	5	1	0.35,0.65	24
Forced response	N	σ^2	β	g	p_1, p_2	n
	60	2	5	1.5	0.35,0.30	53
	80	2	5	1	0.35,0.30	68
Eriksson's RRT	N	σ^2	β	g	a, b, L	n
	60	2	5	1.5	24,-5,33	1699
	80	2	5	1	24,-5,33	1035

Table 4.11 Fixing sample-size for Horvitz-Thompson estimate in RR surveys with $\alpha = 0.05$, $f = 0.1$

Warner's RR	N	σ^2	β	g	p	n
	60	2	15	1.5	0.35	54
	80	2	15	1	0.35	67
URL RRT	N	σ^2	β	g	p_1, p_2	n
	60	2	5	1.5	0.35,0.65	1758
	80	2	5	1	0.35,0.65	1982
Kuk (taking $k = 3$)	N	σ^2	β	g	θ_1, θ_2	n
	60	2	5	1.5	0.35,0.65	39
	80	2	5	1	0.35,0.65	46
Forced response	N	σ^2	β	g	p_1, p_2	n
	60	2	5	1.5	0.35,0.30	44
	80	2	5	1	0.35,0.30	63
Eriksson's RRT	N	σ^2	β	g	a, b, L	n
	60	2	5	1.5	24,− 5,33	212
	80	2	5	1	24,− 5,33	409

Table 4.12 Fixing sample-size for RHC strategy in RR surveys with $\alpha = 0.05$, $f = 0.1$

Warner's RR	N	σ^2	β	g	p	n
	60	2	15	1.5	0.35	13
	80	2	15	1	0.35	13
URL RRT	N	σ^2	β	g	p_1, p_2	n
	60	2	5	1.5	0.35,0.65	12
	80	2	5	1	0.35,0.65	14
Kuk (taking $k = 3$)	N	σ^2	β	g	θ_1, θ_2	n
	60	2	5	1.5	0.35,0.65	13
	80	2	5	1	0.35,0.65	12
Forced response	N	σ^2	β	g	p_1, p_2	n
	60	2	5	1.5	0.35,0.30	13
	80	2	5	1	0.35,0.30	14
Eriksson's RRT	N	σ^2	β	g	a, b, L	n
	60	2	5	1.5	24,− 5,33	12
	80	2	5	1	24,− 5,33	14

$$E_m V(e_{HH}|URL) = E_m V(t_{HH}) + \frac{(1 - p_1)(1 - p_2)(p_1 + p_2 - 2p_1 p_2)}{(p_1 - p_2)^2}$$

$$\frac{1}{n} \sum_{i=1}^{N} \frac{1}{p_i} E_m(y_i - x_i)^2$$

$$= \frac{\sigma^2}{n} \left[X \sum_{i=1}^{N} x_i^{g-1} - \sum_{i=1}^{N} x_i^g \right]$$

$$+ \frac{(1 - p_1)(1 - p_2)(p_1 + p_2 - 2p_1 p_2)}{(p_1 - p_2)^2}$$

$$\frac{1}{n} \sum_{i=1}^{N} \frac{1}{p_i} \left[(\beta - 1)^2 x_i^2 + \sigma^2 x_i^g \right]$$

$$E_m V(e_{HH}|Kuk) = E_m V(t_{HH}) + \frac{1}{n} \sum_{i=1}^{N} \frac{1}{p_i} [b_i(k)\beta x_i + c_i(k)]$$

$$E_m V(e_{HH}|Forced\ Response) = E_m V(t_{HH})$$

$$+ \frac{1}{n} \sum_{i=1}^{N} \frac{1}{p_i} \left[\frac{p_1(1 - p_1) + (1 - p_1 - p_2)(p_2 - p_1)\beta x_i}{(1 - p_1 - p_2)^2} \right]$$

$$E_m V(e_{HH}|Eriksson) = E_m V(t_{HH})$$

$$+ E_m \left[\frac{1}{n} \sum_{i=1}^{N} \frac{1}{p_i} \{a\beta x_i^2 + a\sigma^2 x_i^g + b\beta x_i + L\} \right]$$

$$= \beta^2 X^2 + \sigma^2 \sum_{i=1}^{N} x_i^g + \frac{1}{n} \left[a \left\{ \beta \sum_{i=1}^{N} \frac{1}{p_i} x_i^2 + \sigma^2 \sum_{i=1}^{N} \frac{1}{p_i} x_i^g \right\} \right.$$

$$+ b \sum_{i=1}^{N} \frac{1}{p_i} x_i + L \Bigg]$$

$$E_m(Y^2) = \beta^2 X^2 + \sigma^2 \sum_{i=1}^{N} x_i^g.$$

Fixing sample-size in IPPS sampling and Horvitz-Thompson estimator in RR surveys

$$E_m V(e_{HT}|Warner) = E_m V(t_{HT}) + \sum_{i=1}^{N} \frac{1}{np_i} \frac{p(1 - p)}{(2p - 1)^2}$$

$$E_m V(e_{HT}|URL)$$

$$= E_m V(t_{HT}) + \sum_{i=1}^{N} \frac{1}{np_i}$$

$$\left\{\frac{(1-p_1)(1-p_2)(p_1+p_2-2p_1p_2)}{(p_1-p_2)^2}\right\}$$
$$\{(\beta-1)^2x_i^2+\sigma^2x_i^g\}$$

$$E_m V(e_{\text{HT}}|\text{Kuk}) = E_m V(t_{\text{HT}}) + \frac{1}{n}\sum_{i=1}^N \frac{1}{p_i}\{b_i(k)\beta x_i + c_i(k)\}$$

$$E_m V(e_{\text{HT}}|\text{Forced Response})$$

$$= E_m V(t_{\text{HT}}) + \sum_{i=1}^N \frac{1}{np_i}$$

$$\left\{\frac{p_1(1-p_1)+(1-p_1-p_2)(p_2-p_1)\beta x_i}{(1-p_1-p_2)^2}\right\}$$

$$E_m V(e_{\text{HT}}|\text{Eriksson's RRT}) = E_m V(t_{\text{HT}}) + \sum_{i=1}^N \frac{1}{np_i}\{a\beta x_i^2 + a\sigma^2 x_i^g + b\beta x_i + L\}.$$

Fixing sample-size in RR surveys by RHC strategy in estimating finite population total

$$E_m V(e_{\text{RHC}}) = E_m V(t_{\text{RHC}}) + \sum_{i=1}^N \frac{Q_i}{p_i}V_R(r_i)$$

4.5 Recommendation and Conclusion

The effectiveness of Chebyshev inequality-based sample-size fixing in direct response surveys is supported by numerical evidence. The same, however, does not work to randomized response surveys with Hansen-Hurwitz and Horvitz-Thompson estimator. This may be due to the fact that one term in the variance of an unbiased estimator based on RR data is exclusively determined by the sample design and DR survey data, whereas the other term depends on RR-based variance, which is too high regardless of sampling design details. Therefore, the variance formula's term cannot be effectively controlled by the Chebyshev rule.

Therefore, we advise using Chebyshev inequality to reduce the variance of a component in the variance that exclusively uses direct response-based characteristics. The other variance term cannot be controlled by our approach which aims at controlling only the DR-related materials.

References

Brewer, K. R., & Hanif, M. (1983). *Sampling with unequal probabilities*. Springer.

Chaudhuri, A. (2020). A review on issues of settling the sample-size in surveys: Two approaches—Equal and varying probability sampling—Crises in sensitive cases. *CSA Bulletin, 72*(1), 7–16.

Chaudhuri, A. (2010). *Essentials of survey sampling*. Prentice Hall of India.

Chaudhuri, A. (2011). *Randomized response and indirect questioning techniques in surveys*. CRC Press.

Chaudhuri, A. (2014). *Modern survey sampling*. Chapman & Hall, CRC, Taylor & Francis.

Chaudhuri, A. (2018). *Survey sampling*. Taylor & Francis.

Chaudhuri, A., & Dutta, T. (2018). Determining the size of a sample to take from a finite population. *Statistics and Applications, 16*(1), 37–44.

Chaudhuri, A., & Patra, D. (2023a). Fixing the size of a sample to draw in a randomized response survey. *Pakistan Journal of Statistics, 39*(3), 289–298.

Chaudhuri, A., & Patra, D. (2023b). Fixing size of varying probability sample in a direct and a randomized response survey. *Journal of Indian Society of Agricultural Statistics, 77*(1), 19–26.

Chaudhuri, A., & Sen, A. (2020). Fixing the sample-size in direct and randomized response surveys. *Journal of Indian Society of Agricultural Statistics, 74*(3), 201–208.

Chaudhuri, A., & Vos, J. W. (1988). *Unified theory and strategies of survey sampling*. North-Holland.

Eriksson, S. C. (1973). A new model for RR. *International Statistical Review, 41*, 101–113.

Hansen, M. M., & Hurwitz, W. N. (1943). On the theory of sampling from finite populations. *Annals of Mathematical Statistics, 14*, 333–362.

Kuk, A. Y. (1990). Asking sensitive questions indirectly. *Biometrika, 77*(2), 436–438.

Warner, S. L. (1965). Randomized response: A survey technique for eliminating evasive answer bias. *Journal of American Statistical Association, 60*, 63–69.

Chapter 5
Likelihood Approach and Its Ramifications

5.1 Introduction

Suppose that every person in a population belongs to either sensitive group A or the complement group. It is required to estimate by survey the proportion belonging to Group A. Individuals are reluctant to report their truthful status due to the sensitivity of the survey questions. Warner (1965) suggested an alternate method for increasing cooperation in surveys. A randomized response method for estimating a population proportion was developed to protect the respondent's privacy. Under the assumption that yes and no reports are made truthfully, Maximum Likelihood Estimates (MLEs) of the true population proportion are straightforward. Singh (1976) noted that MLE provided by Warner (1965) is not the MLE of the population proportion and suggested revised MLE.

O'Hagan (1987) derived linear Bayes estimator for the population proportion. Winkler and Franklin (1979) presented a Bayesian approach to Warner's model by postulating a Beta prior probability distribution. Pitz (1980) extended this method to Simmons's (1969) RR model. Barabesi and Marcheselli (2006, 2010) modified Bayesian approach in other RR models with different prior probabilities where samples were drawn by simple random sampling with replacement.

Chaudhuri and Christofides (2013) illustrated a new procedure to estimate a finite population proportion of a sensitive characteristic using Warner's RR data by applying Empirical Bayes (EB) approach when the units are sampled from a finite population using a general sampling scheme. Chaudhuri and Pal (2015) presented logistic regression modelling for different RR models using Empirical Bayesian approach. Chaudhuri and Shaw (2017) employed EB approach in different RR models in the light of Chaudhuri and Christofides (2013). In the present chapter, we estimate population parameter both for qualitative and quantitative data with the above-mentioned different approaches.

5.2   Maximum Likelihood Estimation (MLE)

5.2.1   MLE in Warner's Model

Consider y_i takes value 1 if the $i^{\text{th}}(i = 1, 2, \ldots, N)$ individual in the population U bears a sensitive characteristic A and value 0 if individual i bears A^C. The proportion of individuals $\theta = \frac{1}{N} \sum_{i=1}^{N} y_i$ in the population bearing A is estimated by employing randomized response (RR) devices pioneered by Warner (1965) followed by several other devices existing in literature.

According to Warner (1965), let a simple random sample (SRS) of n people be drawn with replacement (WR) from the population. A sampled person i is offered a box in which $p\left(0 < p < 1, p \neq \frac{1}{2}\right)$ proportion of cards are marked A and the rest of the cards are marked A^C. He/she is requested to draw a card and report yes or no according to his/her match with the sensitive feature A.

Let θ be the probability of bearing the sensitive character A.
Here, $\theta = \frac{1}{N} \sum_{i=1}^{N} y_i, 0 < \theta < 1$.

$$\text{Let} \quad I_i = \begin{cases} 1, & \text{if the } i\text{th person reports "yes"} \\ 0, & \text{if he/she reports "no".} \end{cases}$$

Let

$$P(\text{yes response}) = p\theta + (1 - p)(1 - \theta) = \lambda(\text{say})$$
$$\text{and } P(\text{no response}) = p(1 - \theta) + (1 - p)\theta.$$

The number of "yes" responses is n_1 among n sampled persons.
The likelihood for λ may be written as

$$L(\lambda|n_1, n) = \binom{n}{n_1} \lambda^{n_1} (1 - \lambda)^{n - n_1}$$

The log of the likelihood is proportional to

$$n_1 \log(\lambda) + (n - n_1) \log(1 - \lambda).$$

By solving $\frac{\delta \log(L)}{\delta \lambda} = 0$, the maximum likelihood estimate (MLE) of λ can be written as $\hat{\lambda} = \frac{n_1}{n}$.

Expressing $\lambda = p\theta + (1 - p)(1 - \theta)$, the likelihood for θ is

$$L(\theta|n_1, n) \cong [(1 - p) + (2p - 1)\theta]^{n_1} (1 - [(1 - p) + (2p - 1)\theta])^{n - n_1}$$

Then, the MLE of θ is $\hat{\theta} = \frac{n_1}{n(2p-1)} + \frac{p-1}{(2p-1)}$ with $p \neq \frac{1}{2}$

Alternatively, it can be written as $\hat{\theta} = \frac{\hat{\lambda}-(1-p)}{(2p-1)}$, $p \neq \frac{1}{2}$.

The expectation of $\hat{\theta}$ is

$$E(\hat{\theta}) = \frac{1}{(2p-1)}\left[p - 1 + \frac{1}{n}\sum_{1}^{n}E(y_i)\right]$$

$$= \frac{1}{(2p-1)}[p - 1 + p\theta + (1-p)(1-\theta)] = \theta.$$

So here the MLE $\hat{\theta}$ is unbiased estimate of θ.

The variance is $V(\hat{\theta}) = \frac{nV(y_i)}{n^2(2p-1)^2}$.

On simplification, the variance $V(\hat{\theta})$ can be expressed as the sum of the variance due to sampling and the variance due to the RR device.

5.2.2 Revised MLE

Singh (1976) noted that $\hat{\theta}$ is not the MLE of θ as the MLE must be a member of the parameter space.

The revised MLE is

(i) If $p > \frac{1}{2}$

$$\hat{\lambda} = \begin{cases} \frac{n_1}{n} & \text{if } 1 - p \leq \frac{n_1}{n} < p \\ 1 - p & \text{if } \frac{n_1}{n} < (1-p) \\ p & \text{if } \frac{n_1}{n} > 1 - p. \end{cases}$$

(ii) If $p < \frac{1}{2}$

$$\hat{\lambda} = \begin{cases} \frac{n_1}{n} & \text{if } p < \frac{n_1}{n} \leq (1-p) \\ p & \text{if } \frac{n_1}{n} < p \\ 1 - p & \text{if } \frac{n_1}{n} > (1-p). \end{cases}$$

So the true MLE of θ is $\tilde{\theta}$ with

$$\tilde{\theta} = 1 \quad \text{if} \quad \hat{\lambda} \geq p \geq \frac{1}{2}$$

$$= 0 \quad \text{if} \quad \hat{\lambda} \leq 1 - p < \frac{1}{2}$$

$$= \hat{\theta} \quad \text{if} \quad 1 - p < \hat{\lambda} < (p), p > \frac{1}{2}$$

$$= \hat{\theta} \ \text{ if } \ p < \hat{\lambda} < (1-p), p < \frac{1}{2}$$

$$= 0 \ \text{ if } \ \hat{\lambda} \ge (1-p) > \frac{1}{2}.$$

The expectation and variance of the estimate $\hat{\theta}$ are

$$E\left(\hat{\theta}\right) = \frac{1}{2p-1}\left[(p-1) + \frac{1}{n}\sum_{i \in s} y_i\right]$$

$$= \frac{1}{2p-1}[(p-1) + p\theta + (1-p)(1-\theta)] = \theta,$$

and

$$V\left(\hat{\theta}\right) = \frac{1}{(2p-1)^2}\left[\frac{nV(y_i)}{n^2}\right]$$

$$= \frac{1}{(2p-1)^2}\left[\frac{\{p\theta + (1-p)(1-\theta)\}\{p(1-\theta) + (1-p)\theta\}}{n}\right]$$

$$= \frac{\frac{1}{4} + \{p\theta + (1-p)(1-\theta)\}\}}{(2p-1)^2 n}.$$

Writing $\lambda = \log\left(\frac{\theta}{\bar{\theta}}\right)$, as logistic transformation of θ, Raghavarao (1978) suggested a new estimator of θ to shrink the Warner estimator into admissible range [0,1].

5.2.3 MLE in RR Surveys with Varying Probability Sampling: Chaudhuri's Approach

In a social survey, an unequal probability sample is supposedly at hand. Chaudhuri (2001) proposed an alternative estimation procedure for the proportion θ.

Let a sample s of size n be drawn by general sampling design with probability $p(s)$.

On request, selected person draws randomly one card from the box to respond. Let,

$$I_i = 1 \text{ if card type matches his/her characteristic } A \text{ or } A^C$$

$$= 0, \text{ otherwise.}$$

The person's true value is y_i which is either 1 or 0.
So, Prob$(I_i = 1) = py_i + (1-p)(1-y_i)$.

The likelihood of y_i given I_i is

$$L(y_i|I_i) = [p^{y_i}(1-p)^{1-y_i}]^{I_i}[p^{1-y_i}(1-p)^{y_i}]^{1-I_i}.$$

The MLE of y_i based on Warner's RRT under general sampling scheme is

$$\widehat{y_i} = I_i \text{ if } p > \frac{1}{2}$$

$$= 1 - I_i \text{ if } p < \frac{1}{2}$$

For $p = \frac{1}{2}$, MLE of y_i does not exist. Details are given in Chaudhuri's (2011) monograph.

This RR-based expectation and variance operators generically as E_R, V_R give

$$E_R(I_i) = py_i + (1-p)(1-y_i), \ y_i = 1(0) \text{ if } i \text{ bears A } (A^c)$$
$$V_R(I_i) = p(1-p) \text{ since } I_i = 1/0 \text{ and } y_i = 1/0.$$

Then,

$$r_i = \frac{I_i - (1-p)}{2p-1} \text{ has } E_R(r_i) = y_i \text{ and}$$

$$V_i = V_R(r_i) = \frac{p(1-p)}{(2p-1)^2}, \text{ a known number.}$$

Writing the likelihood as given earlier, the maximum likelihood estimators can be obtained for Unrelated Question RR model, Kuk's model, Mangat and Singh's RR models. Details of the derivations are narrated in Chaudhuri (2011, pp 67–73).

5.3 Bayes Estimation

5.3.1 Bayes Linear Estimators for RR Models

Bayes linear estimators provide simple Bayesian methods and require a minimum of prior specification. O'Hagan (1987) proposed Bayes linear estimators for a variety of randomized response models. In RR, Bayesian methods are attractive because they permit the incorporation of potentially useful prior information. To construct the Bayes linear estimator, prior belief about the $y_i's (i = 1, 2, \ldots, N)$ is needed. It is assumed that every y_i has the same prior mean and variance and the (y_i, y_j) has the same prior covariance. Exchangeability is assumed.

The population proportion that has the sensitive attribute is $\theta = \frac{1}{N}\sum_1^N y_i$.

Let first n individuals of the population be sampled.

The simple random sample from the population consists of n responses. The term $x_j = 1$ if the jth response is "yes", otherwise $x_j = 0$. The expectation, variance of y_i and covariance of y_i, y_j are the following:

$$\text{Let } E(y_i) = m, \; V(y_i) = m(1 - m), \text{cov}(y_i, y_j) = rm(1 - m)$$
$$i, j = 1, 2, \ldots, N$$

and $r \in [-(N - 1)^{-1}, 1]$, the correlation of any pair (y_i, y_j).

The prior expectations are

$$E(\theta) = m,$$
$$V(\theta) = \frac{m(1 - m)[1 + r(N - 1)]}{N},$$
$$E(x_i) = pm + (1 - p)(1 - m) = (1 - p) + (2p - 1)m,$$
$$V(x_i) = p(1 - p) + m(1 - m)(2p - 1)^2,$$
$$\text{Cov}(x_i x_{j.}) = (2p - 1)^2 rm(1 - m),$$

and

$$\text{Cov}(\theta, x_{i.}) = (2p - 1)m(1 - m)[1 + r(N - 1)]/N$$

Drawing a simple random sample of size n from a population of N units and defining $\bar{x} = \frac{1}{n}\sum_1^n x_i$, the first two moments can be written as

$$E(\bar{x}) = (1 - p) + m(1 - 2p),$$
$$V(\bar{x}) = \frac{1}{n}[p(1 - p) + m(1 - m)(2p - 1)^2\{1 + r(n - 1)\}],$$
$$\text{Cov}(\theta, \bar{x}) = (2p - 1)^2 rm(1 - m),$$
$$\hat{\theta}(\bar{x}) = m + \frac{V(\bar{x})}{Cov(\theta, \bar{x})}[\bar{x} - (1 - p) - (2p - 1)m].$$

It can be shown that the above estimate a convex linear combination of Bayesian and non-Bayesian estimates.

The posterior variance is

$$\hat{\theta}(\bar{x}) = \left(1 - (2p - 1)\frac{Cov(\theta, \bar{x})}{V(\bar{x})}\right)\frac{m(1 - m)\left[1 + \frac{r(N-1)}{N}\right]}{N}.$$

5.3.2 *Warner's RR Model: A Bayesian Approach*

Winkler and Franklin (1979) concluded that instead of a maximum likelihood estimator or an unbiased estimator, a Bayesian estimator for θ can be proposed.

Postulating beta distribution as a prior distribution of θ and likelihood as binomial distribution, the posterior distribution can be written.

Let the prior distribution be a beta (α, β) distribution which is

$$g(\theta|\alpha, \beta) = \frac{1}{B(\alpha, \beta)} \theta^{\alpha-1}(1-\theta)^{\beta-1}, \alpha > 0, \beta > 0,$$

where $B(\alpha, \beta)$ is the beta function.

The likelihood function for the above model is

$$[(1-p) + (2p-1)\theta]^{n_1} (1 - [(1-p) + (2p-1)\theta])^{n-n_1}.$$

So, the posterior density is

$$h(\theta|n_1, n) \propto [(1-p) + (2p-1)\theta]^{n_1}$$
$$(1 - [(1-p) + (2p-1)\theta])^{n-n_1} \theta^{\alpha-1}(1-\theta)^{\beta-1}$$

for $0 < \theta < 1$.

Expanding the terms in square brackets and normalizing, it can be written as a linear combination of several beta densities

$$h(\theta|n_1, n) = \sum_{t=0}^{n} w_t h(\theta|\alpha + t, \beta + n - t),$$

where $w_t = \frac{w_t^*}{\sum_{t=0}^{n} w_t^*}$ and

$$w_t^* = \binom{n}{t} \frac{B(\alpha + t, \beta + n - t)}{B(\alpha, \beta)} \sum_{j=0}^{\min(n_1, t)} \binom{t}{j} \binom{n-t}{n_1 - j} p^{n-t-n_1+2j} (1-p)^{t+n_1-2j}.$$

So the posterior distribution of θ is a mixture distribution.

5.4 Empirical Bayes (EB) Estimation

5.4.1 EB Estimation for Qualitative Data

Chaudhuri and Christofides (2013) in their monograph illustrated a new procedure to estimate a finite population proportion of a sensitive characteristic using Warner's RR data by applying Empirical Bayes approach when the units are sampled from a finite population using a general sampling scheme.

Denoting L_i as prior probability of $y_i = 1$, the posterior probability of $y_i = 1$ given I_i is

$$
\begin{aligned}
L_i(1) &= \frac{pL_i}{pL_i + (1-p)(1-L_i)} \\
\frac{1}{L_i(1)} &= \frac{(1-p) + pL_i - (1-p)(L_i)}{pL_i} \\
&= \frac{(1-p)}{pL_i} + \frac{(2p-1)}{p}
\end{aligned}
\tag{5.1}
$$

which gives $L_i = \frac{(1-p)}{p} \left(\frac{1}{L_i(1)} - \frac{(2p-1)}{p} \right)^{-1}$.

Taking $L_i(1) = r_i$, an estimator of y_i and plugging in Eq. (5.1), an estimator for L_i can be given by

$$
\begin{aligned}
\hat{L}_i &= \frac{(1-p)}{p} \left(\frac{1}{r_i} - \frac{(2p-1)}{p} \right)^{-1} \\
&= \alpha \left(\frac{1}{r_i} - \beta \right)^{-1}, \text{ where } \alpha = \frac{1-p}{p} \text{ and } \beta = \frac{2p-1}{p} \\
&= \alpha \left(\frac{r_i}{1 - r_i \beta} \right) \quad \text{such that } E_R(\hat{L}_i) \cong L_i
\end{aligned}
$$

Expanding $\hat{L}_i (= f(r_i))$ about $L_i (= f(L_i(1)))$ using Taylor's Expansion and neglecting higher order terms we get,

$$
\begin{aligned}
\hat{L}_i &= L_i + \frac{\partial}{\partial r_i} f(r_i)|_{r_i = L_i(1)} (r_i - L_i(1)) \\
&= L_i + \frac{\alpha}{(1 - L_i(1)\beta)^2} (r_i - L_i(1)).
\end{aligned}
$$

So,

$$
V_R(\hat{L}_i) = \frac{\alpha^2}{(1 - L_i(1)\beta)^4} V_R(r_i).
$$

It is estimated as $\hat{V}_R(\hat{L}_i) = \frac{\alpha^2}{(1-r_i\beta)^4}\upsilon_R(\mathsf{r}_i) = \frac{p^3(1-p)^3}{[p-r_i(2p-1)]^4(2p-1)^2} =$

$\upsilon_R(\hat{L}_i).\hat{V}_R(\hat{L}_i) = \frac{\alpha^2}{(1-r_i\beta)^4}\upsilon_R(\mathsf{r}_i) = \frac{p^3(1-p)^3}{[p-r_i(2p-1)]^4(2p-1)^2} = \upsilon_R(\hat{L}_i).$

Proceeding similarly as above,

$$L_i(0) = \text{Posterior Prob}[y_i = 1 | I_i = 0]$$

$$= \frac{P[I_i = 0 | y_i = 1]L_i}{P[I_i = 0 | y_i = 1]L_i + P[I_i = 0 | y_i = 0](1 - L_i)}$$

$$= \frac{(1 - p)L_i}{p + (1 - 2p)L_i}$$

$$L_i = \frac{p}{1 - p}\left(\frac{1}{L_i(0)} - \frac{1 - 2p}{1 - p}\right)^{-1}$$

$$\hat{L}_i = \frac{p}{1 - p}\left(\frac{1}{r_i} - \frac{1 - 2p}{1 - p}\right)^{-1}$$

$$E_R(\hat{L}_i) \cong L_i$$

$$\hat{V}_R(\hat{L}_i) = \frac{p^3(1 - p)^3}{[(1 - p) + r_i(2p - 1)]^4(2p - 1)^2} = \upsilon_R(\hat{L}_i).$$

Now putting the expression for r_i in the expressions of $\hat{L}_i$ and $\hat{V}_R(\hat{L}_i)$, it is seen that $\hat{L}_i$ and $\hat{V}_R(\hat{L}_i)$ do not exist, making numerical calculations impossible.

Then, in place of estimating $\theta = \frac{1}{N}\sum_{i=1}^N y_i$, the term $\overline{L} = \frac{1}{N}\sum_{i=1}^N L_i$ may be estimated by $\widehat{\overline{L}} = \frac{1}{N}\sum_{i\in s}\frac{\hat{L}_i}{\pi_i} = e$ *vide* Chaudhuri (2011) and the variance of $\widehat{\overline{L}}$ can be derived.

5.4.1.1 Logistic Regression Modelling

An improvement by Empirical Bayes procedure using a correlated variable while using a randomized response technique may be employed.

Pal and Chaudhuri (2018) developed Empirical Bayes models with Chaudhuri's (2011) approach on Warner (1965), Boruch's Forced Response (1972), Kuk (1990) and Simmon's URL RRT models. The details of the RR models are described in Chap. 4.

Following Chaudhuri and Saha (2004), a one-parameter logistic regression model postulated for $\theta = \frac{1}{N}\sum_1^N y_i$ $(0 < \theta < 1)$ is

$$\log\left(\frac{\theta}{1 - \theta}\right) = \psi\left(\underset{\sim}{x}\right), \underset{\sim}{x} = (x_1, \ldots, x_i, \ldots, x_N).$$

The model can be assumed as

$$\log\left(\frac{\hat{\theta}}{1-\hat{\theta}}\right) = \log\left(\frac{\theta}{1-\theta}\right) + \epsilon$$

where $\hat{\theta}$ is the estimator for θ with $0 < \hat{\theta} < 1$. A suitable probability distribution is assumed for the error term ϵ. They pointed out the inapplicability of this approach for improved estimation of θ to

(a) Warner's RRT and (b) Forced Response RRT of Boruch (1972) because the requisite conditions on θ, $\hat{\theta}$ (generically) do not hold.

Chaudhuri and Pal (2015) pointed out these issues and provided the modifications as follows:

(I) **Warner's RRT**:

Let

$$\theta_i = (2p - 1)y_i + p(1 - y_i).$$

It is estimated as

$$\hat{\theta}_i = (2p - 1)r_i + p(1 - r_i).$$

Then,

$$\hat{\theta}_i\big|_{I_i=1} = \frac{3p^2 - 2p}{2p - 1} > 0 \;\; \text{if} \;\; p > \frac{2}{3}$$
$$< 1 \; \text{if} \; 3p^2 - 4p + 1 < 0.$$

Also,

$$\hat{\theta}_i\big|_{I_i=0} = \frac{3p^2 - 2p + 1}{2p - 1} > 0 \;\; \text{if} \;\; p(1 - p) > \frac{1}{3}$$
$$< 1 \;\; \text{if} \; 3p^2 - 5p + 2 < 0.$$

Now the variance term is

$$V\left(\hat{\theta}_i\right) = (1 - p)^2 V_i > 0.$$

On choosing p subject to the conditions, it is ensured that $0 < \hat{\theta}_i < 1 \; \forall i$.

(ii) Similarly, for **Forced Response RRT**, let

$$\theta_i = (1 - p_1 - p_2)y_i + (1 - p_1)(1 - y_i)$$

and

$$\hat{\theta}_i = (1 - p_1 - p_2)r_i + (1 - p_1)(1 - r_i).$$

Then,

$$\hat{\theta}_i\big|_{I_i=1} = \frac{(1 - p_1)^2}{1 - p_1 - p_2} > 0$$

and

$$< 1 \text{ if } p_1(1 - p_1) > p_2$$

and

$$\hat{\theta}_i\big|_{I_i=0} = \frac{(1 - p_1)(1 - p_1 - p_2) + p_1 p_2}{(1 - p_1 - p_2)} > 0$$

and

$$< 1 \text{ if } (1 - p_1)^2 - p_2(1 - p_1) + p_1 p_2 < (1 - p_1 - p_2)$$

with appropriate p_1, p_2 both θ_i and $\hat{\theta}_i$ take values in the open interval $(0,1)$.

Also, $V\left(\hat{\theta}_i\right) = p_2^2 V_i \forall i$.

Though Kuk's RRT and URL of Simmons did not pose to Chaudhuri and Saha (2004) any problem for use of one-parameter logistic regression modelling let us still note.

(iii) **Kuk's (1990) RRT**:

Under Kuk's model, we may write

$$\theta_i = (p_1 - p_2)y_i + p_2,$$
$$\hat{\theta}_i = (p_1 - p_2)r_i + p_2.$$

Both belong to $(0,1)$ for $k > 0$,

$$V\left(\hat{\theta}_i\right) = (p_1 - p_2)^2 V_i$$
$$= (p_1 - p_2)^2 (a_i + b_i) \text{ if } y_i = 1$$
$$= (p_1 - p_2)^2 b_i \text{ if } y_i = 0$$

$$\text{with } a_i = \frac{(1 - p_1 - p_2)}{k(p_1 - p_2)^2} \text{ and } b_i = \frac{p_2(1 - p_2)}{k(p_1 - p_2)^2}.$$

(iv) **URL Model of Simmons**:

Let $\hat{\theta}_i = \frac{p_1 - p_2}{2 - (p_1 + p_2)}\left(r_i + \frac{1 - p_1}{p_1 - p_2}\right)$ and $\theta_i = \hat{\theta}_i\big|_{r_i = y_i}$.

Both $\hat{\theta}_i$, θ_i are points in (0,1).

For simplicity, we may write

$$\text{logit}(\theta_i) = \log\left(\frac{\theta_i}{1 - \theta_i}\right).$$

It is modelled as

$$\text{logit}(\hat{\theta}_i) = \text{logit}(\theta_i) + e_i, \; e_i \sim N(0, \hat{V}_i(\text{logit}).$$

" $\sim$ " means distributed independently.

The variance is

$$V_i(\text{logit}) = V\left[\text{logit}(\hat{\theta}_i)\right] = V\left[\log\left(\frac{\hat{\theta}_i}{1 - \hat{\theta}_i}\right)\right] = \frac{V(\hat{\theta}_i)}{\theta_i(1 - \theta_i)^2}$$

and $\widehat{V}_i(\text{logit}) = V_i(\text{logit})|_{\theta_i = \hat{\theta}_i}, \; \forall i.$

Let, further, the model satisfy

$$\text{logit}(\theta_i) \sim N\left(x_i \beta_{\text{logit}}, A_{\text{logit}}\right).$$

Here, β_{logit} is an unknown constant, and A_{logit} is another unknown constant.

Now β_{logit} and A_{logit} will be estimated iteratively following Fay and Herriot's (1979) procedure as employed in small area estimation context. Letting

$$\hat{\beta}_{\text{logit}} = \frac{\sum_{i \in s} \text{logit}(\hat{\theta}_i) x_i / \left(A_{\text{logit}} + \hat{V}_i(\text{logit})\right)}{\sum_{i \in s} 1 / \left(A_{\text{logit}} + \hat{V}_i(\text{logit})\right)} \tag{5.2}$$

and noting

$$\sum \frac{\left[\text{logit}(\hat{\theta}_i) - \hat{\beta}_{\text{logit}}\right]^2}{A_{\text{logit}} + \hat{V}_i(\text{logit})} \tag{5.3}$$

is distributed as a chi-square variable with degrees of freedom equal to $(n-1)$ and supposing n as the sample-size, it is easy to estimate by iteration A_{logit} and β_{logit} by method of moments solving Eq. (5.3) $= (n - 1)$. Let these estimates be $\hat{A}_{\text{logit}}$ and $\hat{\beta}_{\text{logit}}$.

Then, $logit(\theta_i)$ may be estimated by the Empirical Bayes estimate

$$logit(\hat{\theta}_{i(EB)}) = \left(\frac{\hat{A}_{logit}}{\hat{A}_{logit} + \hat{V}_{i\,logit}}\right) logit(\hat{\theta}_i) + \left(\frac{\hat{V}_{i\,logit}}{\hat{A}_{logit} + \hat{V}_{i\,logit}}\right)\beta_{logit}x_i$$

$$= \lambda_i, \text{ say}$$

Then, θ_i is estimated by

$$\hat{\theta}_{i(EB)} = \frac{e^{\lambda_i}}{1 + e^{\lambda_i}}.$$

We follow Prasad and Rao (1990) to estimate the mean square error (MSE) of $logit(\hat{\theta}_{iEB})$ in the following way:

Let

$$g_{1i} = (1 - logit(\theta_{iEB}))\hat{V}_{i\,logit},$$

$$g_{2i} = \left[logit(\hat{\theta}_{iEB})\right]^2 \frac{x_i^2}{\sum_1^n \left(\frac{x_i^2}{\hat{A}_{logit} + \hat{V}_{i\,logit}}\right)},$$

$$g_{3i} = \frac{(\hat{V}_{i\,logit})^2}{(\hat{A}_{logit} + \hat{V}_{i\,logit})^3} \frac{2}{\sum_1^n \left(\frac{1}{\hat{A}_{logit} + \hat{V}_{i\,logit}}\right)}.$$

Then, the MSE of $logit(\hat{\theta}_{iEB}) = \lambda_i$ is estimated as

$$m_i = g_{1i} + g_{2i} + 2g_{3i}.$$

Then, MSE of $\hat{\theta}_{i(EB)}$ is estimated on noting the well-known formula

$$V[f(\hat{t}(x))] \approx \left[\frac{\partial f(\hat{t}(x))}{\partial x}\right]^2 V(\hat{t}(x)).$$

If $t(x)$ is estimated by $\hat{t}(x)$, then $V[f(\hat{t}(x)]$ is estimated by $V[f(t(x)]|_{t(x)=\hat{t}(x)}$. Empirical Bayes estimator for Warner's (1965) model is the following:

$$r_{i(logit)} = \frac{(p - \hat{\theta}_{i(EB)})}{1 - p} \text{ has } E_R(r_{i(logit)}) = y_i \text{ and}$$

$$V_{i(logit)} = V_R(r_{i(logit)}) = \frac{m_i}{(1-p)^2}, m_i \text{ is the estimated MSE of } logit(\hat{\theta}_{iEB}) = \lambda_i.$$

Let E_P, V_P denote the design-based operator for expectation, variance and E_M, V_M are the model-based operator for expectation and variance.

Let $E = E_P E_R E_M$ and $= E_P E_R V_M + E_P V_R E_M + V_P E_R E_M$

Then, it may be noted that $E_P V_R E_M(e_{\text{EB}}) = 0$.

Here considering Horvitz-Thompson (HT) estimator, the estimator for θ is

$$e_{\text{EB}} = \frac{1}{N} \sum_{i \in s} \frac{r_{i(\text{logit})}}{\pi_i}$$

$$E(e_{\text{EB}}) = E_P E_R E_M(e_{\text{EB}}) = \frac{1}{N} E_P E_R \left(\frac{p - \hat{\theta}_{i(\text{EB})}}{(1-p)\pi_i} \right)$$

$$= \frac{1}{N} E_P \sum_{i \in s} \frac{p - (2p-1)y_i - p(1-y_i)}{(1-p)\pi_i} = \frac{1}{N} E_P \left(\sum_{i \in s} \frac{y_i}{\pi_i} \right) = \theta.$$

The related variance term

$$V(e_{\text{EB}}) = E_P E_R V_M(e_{\text{EB}}) + E_P V_R E_M(e_{\text{EB}}) + V_P E_R E_M(e_{\text{EB}}).$$

Here, $E_P E_R V_M(e_{\text{EB}}) = \frac{1}{N^2(1-p)^2} E_P E_R \sum_{i \in s} \frac{M_i}{\pi_i^2} = \frac{1}{N^2(1-p)^2} E_P \sum_1^N \frac{M_i}{\pi_i^2} = \sum_1^N \frac{M_i}{\pi_i}$, where M_i is MSE of $\text{logit}(\hat{\theta}_{i\text{EB}})$.

Now, $E_P V_R E_M(e_{\text{EB}}) = E_P V_R \sum_{i \in s} \frac{p - \theta_i}{N\pi_i} = 0$.

$$V_P E_R E_M(e_{\text{EB}}) = \frac{1}{N^2} V_P E_R \sum_{i \in s} \frac{p - \theta_i}{\pi_i(1-p)} = \frac{1}{N^2} V_P \left(\sum_{i \in s} \frac{y_i}{\pi_i} \right)$$

$$= \frac{1}{N^2} \sum \sum_{i<j} (\pi_i \pi_j - \pi_{ij}) \left(\frac{y_i}{\pi_i} - \frac{y_j}{\pi_j} \right)^2$$

The variance $V(e_{EB})$ may be estimated by

$$m(e_{\text{EB}}) = \frac{1}{N^2(1-p)^2} \sum_{i \in s} \frac{m_i}{\pi_i^2} + \frac{1}{N^2} \left[A - \sum \sum_{i<j \in s} (\pi_i \pi_j - \pi_{ij}) \left(\frac{m_i}{\pi_i} - \frac{m_j}{\pi_j} \right)^2 \frac{1}{\pi_{ij}} \right],$$

where $A = \sum_{i<j \in s} \sum (\pi_i \pi_j - \pi_{ij}) \left(\frac{r_{i(\text{logit})}}{\pi_i} - \frac{r_{j(\text{logit})}}{\pi_j} \right)^2 \frac{1}{\pi_{ij}}$.

5.4.1.2 Unknowable Prior Probabilities

From (5.1), we get

$$L_i = \frac{(1-p)}{p} \left(\frac{1}{L_i(1)} - \frac{(2p-1)}{p} \right)^{-1}$$

where L_i be the prior Probability of $y_i = 1$.

$$L_i(1) = \text{Posterior Prob}[y_i = 1 | I_i = 1]$$

As an estimator of $L_i(1)$, an unbiased estimator of y_i is taken.

Adopting Warner's RR model in general sampling design we may proceed as follows:

Let a box containing similar cards marked A and A^c in proportions $p(\neq 0.5)$ and $(1-p)$ respectively is provided to a sampled person i. The participant's response is

$I_i = 1$ if the card type drawn by i matches his/her characteristic

$= 0$ if there is no match.

Then, $r_i = \frac{I_i-(1-p)}{2p-1}$

is an unbiased estimator of y_i such that $E_R(r_i) = y_i$ and

$$V_R(r_i) = \frac{p(1-p)}{(2p-1)^2} \text{ is } v_R(r_i).$$

Any RR model like Kuk's (1990), Christofides' (2003), Mangat and Singh's (1990), Forced Response and Unrelated Question RR model can be used to estimate y_i. The different RR models are narrated in Chap. 4, and the estimators of y_i are given.

Proceeding as described in Sect. 5.4.1, the final Bayes estimators along with their variance estimators can be calculated. Shaw and Chaudhuri (2016) examined how the EB method may work out with other RRT's like Kuk's (1990), Christofides' (2003), Mangat and Singh's(1990), Forced Response and Unrelated Question models.

5.4.2 EB Estimation for Quantitative Data

5.4.2.1 With Regression Modelling

In this section, the Empirical Bayes (EB) approach in estimating the population mean of a sensitive quantitative variable is described.

Suppose y_i denotes income earned by clandestine means, or the cost of AIDS treatment, or some other stigmatizing value for the ith member of a survey population $U = (1, 2, \ldots N)$, $i = (1, 2, \ldots N)$.

Here, the problem is to estimate the mean $\theta = \frac{1}{N} \sum_{i=1}^{N} y_i$.

Here, Chaudhuri and Pal (2015) used EB approach with logistic regression modelling in estimating the population mean of a sensitive quantitative variable. At first they considered the Device I vide Chaudhuri (2011) in which a person i is offered two boxes with identical cards marked $a_1, a_2, \ldots, a_T$ with nonzero mean $\mu_a(\neq 0)$ and variance σ_a^2 in sufficient numbers placed in first box and those numbered $b_1, b_2, \ldots, b_M$ in second box with mean and variance μ_b, and σ_b^2. Then, i has to draw one card independently from each of the two boxes, say, bearing a_j and b_k, and report $z_i = a_j y_i + b_k$ to the investigator without disclosing any card.

Here, $E_R(z_i) = \mu_a y_i + \mu_b$. So y_i can be unbiasedly estimated as $r_i = \frac{z_i - \mu_b}{\mu_a}$.

$$V_R(r_i) = \frac{\sigma_a^2}{\mu_a^2} y_i^2 + \frac{\sigma_b^2}{\mu_a^2}.$$

Considering a suitable variable x which is correlated with y, Chaudhuri and Pal (2015) suggested an Empirical Bayes approach as an alternative estimation procedure. A suitable model on r_i can be written by $r_i = y_i + \epsilon_i$ such that all ϵ_i are distributed independently and normally, $\epsilon_i \sim N(0, v_i), i \in U$.

Also assume $y_i \sim N(\beta x_i, A), i \in U$. Proceeding as in Fay-Herriot (1979) model of small area estimation approach, the conditional distribution of r_i given y_i is $r_i | y_i \sim N(y_i, v_i)$, and marginally $r_i \sim N(\beta x_i, A + v_i), i = 1, 2, ..N$.

The joint distribution of r_i and y_i is bivariate normal.

Here,

$$\begin{pmatrix} r_i \\ y_i \end{pmatrix} \sim N\left(\begin{pmatrix} r_i \\ y_i \end{pmatrix}, \begin{pmatrix} A + v_i & A \\ A & A \end{pmatrix} \right)$$

So the posterior distribution of y_i given r_i is

$$y_i | r_i \sim N\left(\beta x_i + \frac{A}{A + v_i}(r_i - \beta x_i), \frac{A v_i}{A + v_i} \right), i = 1, 2, \ldots, N.$$

So the Bayes estimator is

$$r_i(B) = \frac{A}{A + v_i} r_i + \frac{v_i}{A + v_i} \beta x_i.$$

It is not usable as the model variance A and regression coefficient β are unknown. The coefficient β can be estimated as

$$\tilde{\beta} = \frac{\sum_{i \in s} r_i x_i / (A + v_i)}{\sum_{i s} x_i^2 / (A + v_i)} \tag{5.4}$$

It is noted that $\sum_{i \epsilon s} \frac{(r_i - \tilde{\beta} x_i)^2}{A + v_i}$ is distributed as a chi-square variable with degrees of freedom $\gamma(s)$, the number of distinct units in the sample s.

So we may write

$$\sum_{i \epsilon s} \frac{(r_i - \tilde{\beta} x_i)^2}{A + v_i} = \gamma(s) - 1. \tag{5.5}$$

Starting with an initial value of A as A_0, we get a value of β from Eq. (5.4).

Using the β in Eq. (5.5), the β and A are solved by iteration method. So, solving (5.4) and (5.5) iteratively we can estimate β and A as $\hat{\beta}$ and $\hat{A}$.

So the revised Bayes estimator of y_i is

$$r_i(EB) = \frac{\hat{A}}{\hat{A} + v_i} r_i + \frac{v_i}{\hat{A} + v_i} \hat{\beta} x_i. \tag{5.6}$$

It is termed as Empirical Bayes (EB) estimator of y_i, $i \in s$.

We can write $r_i(EB)$ as a convex combination of model-based estimate $\hat{\beta} x_i$ and sample-based estimate r_i. So

$$r_i(EB) = \left(1 - \hat{B}_{i.}\right) r_i + \hat{B}_i \left(\hat{\beta} x_i\right), \text{ Where } \hat{B}_i = \frac{v_i}{\hat{A} + v_i}.$$

Following the approach of Prasad and Rao (1990) as in small area estimation context, the variance estimate of $r_i(EB)$ is

$$m_i = g_{1i}\left(\hat{A}\right) + g_{2i}\left(\hat{A}\right) + 2 g_{3i}\left(\hat{A}\right),$$

where $g_{1i}\left(\hat{A}\right) = \left(1 - \hat{B}_{i.}\right) v_i$,

$$g_{2i}\left(\hat{A}\right) = \hat{B}_i^2 \frac{x_i^2}{\sum_{i \epsilon s} x_i^2 / (\hat{A} + v_i)}.$$

And

$$g_{3i}\left(\hat{A}\right) = \frac{v_i^2}{\left(\hat{A} + v_i\right)^3} \frac{2}{n^2} \sum_{i \epsilon s} \left(\hat{A} + v_i\right)^2 \quad \text{for } i \in s.$$

5.4.2.2 With Unknown Prior Probability Modelling

Chaudhuri and Shaw (2017) provided Empirical Bayes estimation of the population mean by assigning unknown prior probabilities to the individuals of the finite population.

Let $L_i = L(y_i) = $ Prior Probability

Then, using Chaudhuri and Christofides (2013), we have,

$$L(y_i|z_i) = \text{Posterior Probability}$$
$$= \frac{L_i P(z_i|y_i)}{P(z_i)}$$
$$= \frac{L_i(1/(\text{TM}))}{(1/(\text{TM}))}$$
$$= L_i$$

Now, $E_{L_i^*}(y_i|z_i) = \frac{\sum_{j=1}^{T} \sum_{k=1}^{M} L_i y_i}{\sum_{j=1}^{T} \sum_{k=1}^{M} L_i}$, where $E_{L_i^*}(y_i|z_i)$ is the Empirical Bayes estimate for y_i.

$$= \frac{\sum_{j=1}^{T} \sum_{k=1}^{M} \left(\frac{z_i - b_k}{a_j}\right)}{\text{TM}} \left[\text{since } z_i = a_j y_1 + b_k\right]$$

$$= \frac{z_i}{T} \sum_{j=1}^{T} \frac{1}{a_j} - \frac{\sum_{j=1}^{T} \frac{1}{a_j} \sum_{k=1}^{M} b_k}{\text{TM}}$$

$$= \frac{z_i - \mu_b.}{\text{HM}_a} = h_i(\text{say}), \text{ Where HM}_a = \frac{1}{T} \sum_{j=1}^{T} \frac{1}{a_j}, \text{ the harmonic mean.}$$

The term h_i is the Empirical Bayes estimate for y_i.

$$E_R = (h_i) = E_R\left(\frac{a_j y_i + b_k - \mu_b}{\text{HM}_a}\right) = \frac{\mu_a y_i}{\text{HM}_a}$$

$$V_R(h_i) = \frac{V_R(z_i)}{\text{HM}_a^2}$$

$$= \frac{\sigma_a^2 y_i^2 + \sigma_b^2}{\text{HM}_a^2}$$

$$= \alpha' y_i^2 + \beta'$$

where $\alpha' = \frac{\sigma_a^2}{\text{HM}_a^2}$ and $\beta' = \frac{\sigma_b^2}{\text{HM}_a^2}$.

$$E_R\left[\frac{\alpha' h_i^2 + \beta'}{1 + \alpha'}\right] = \frac{\frac{\sigma_a^2}{\text{HM}_a^2}\left(\frac{\sigma_a^2 y_i^2 + \sigma_b^2 + \mu_a^2 y_i^2}{\text{HM}_a^2}\right) + \frac{\sigma_b^2}{\text{HM}_a^2}}{1 + \frac{\sigma_a^2}{\text{HM}_a^2}}.$$

So, $\frac{\alpha' h_i^2 + \beta'}{1 + \alpha'}$ can be considered as an estimator of $V_R(h_i)$.

We then estimate $\overline{L} = \frac{1}{N} \sum_{i=1}^{N} L_i$ instead of $\overline{Y} = \frac{1}{N} \sum_{i=1}^{N} y_i$ by taking h_i as an initial estimator for $L_i(y_i)$. The estimator for $\overline{L}$ is $\hat{\overline{L}} = \frac{1}{N} \sum_{i \in s} \frac{h_i}{\pi_i}$, vide Chaudhuri (2011) and variance of $\hat{\overline{L}}$ is then derived and estimated.

References

Barabesi, L., & Marcheselli, M. (2006). A practical implementation and Bayesian estimation in Franklin's randomized response procedure. *Communication in Statistics- Simulation and Computation, 35,* 563–573.

Barabesi, L., & Marcheselli, M. (2010). Bayesian estimation of proportion and sensitivity level in randomized response procedures. *Metrika, 72,* 75–88.

Boruch, R. F. (1972). Relations among statistical methods for assuring confidentiality of social reseach data, *Social Science Research, 1,* 403–414.

Chaudhuri, A. (2001). Using a randomized response from a complex survey to estimate a sensitive proportion in a dichotomous finite population, *Journal of Statistical Planning and Inference, 94*(1), 37–42.

Chaudhuri, A. (2011). *Randomized response and indirect questioning techniques in surveys* (Vol. 348). CRC Press.

Christofides, T. C. (2003). A generalized randomized response technique. *Metrika, 57,* 195–200.

Chaudhuri, A., & Christofides, T. C. (2013). *Indirect questioning in sample surveys.* Springer.

Chaudhuri, A., & Pal, S. (2015). On efficacy of empirical Bayes estimation of a finite population mean of a sensitive variable through randomized responses. *Model Assisted Statistics and Applications, 10*(4), 283–288.

Chaudhuri, A., & Shaw, P. (2017). Empirical Bayes estimation using quantitative randomized response data. *Statistics and Applications, New Series, 15*(1,2), 1–6.

Chaudhuri, A., & Saha, A. (2004). Utilizing covariates by logistic regression modeling in improved estimation of population proportions bearing stigmatizing features through randomized responses in complex surveys. *Journal of the Indian Society of Agricultural Statistics, 58,* 190–211.

Fay, R. E., & Herriot, R. A. (1979). Estimates of income for small places: An application of James Stein procedures to census data. *Journal of the American Statistical Association, 74,* 269–277.

Greenberg, B. G., Abul-Ela, A.-L., Simons, W. R., & Horvitz, D. G. (1969). The unrelated question RR model: Theoretical framework. *JASA, 64,* 520–539.

Horvitz., D. G., & Thompson, D. J. (1952). A generalization of sampling without replacement from a finite universe. *Journal of the American Statistical Association, 47,* 663–685.

Kuk, A. Y. C. (1990). Asking sensitive questions indirectly. *Biometrika, 77*(2), 436–438.

Mangat, N. S., & Singh, R. (1990). An alternative randomized response procedure. *Biometrika, 77*(2), 439–442.

O'Hagan, A. (1987). Bayes linear estimators for randomized response models. *Journal of the American Statistical Association, 82*(398), 580–585.

Pal, S., & Chaudhuri, A.(2018). Model assisted estimation of sensitive proportions from randomised responses by unequal probability sampling. *Model Assisted Statistics and Applications, 13*(1), 53–61 .

Pitz, G. F. (1980). Bayesian analysis of random response models. *Psychological Bulletin, 87,* 209–212.

Prasad, N. G. N., & Rao, J. N. K. (1990). The estimation of the mean square error of small area estimates. *Journal of the American Statistical Association, 85,* 163–171.

Raghavarao, D. (1978). On an estimation problem in Warner's randomized response technique. *Biometrics, 34*(1), 87–90.

Shaw, P., & Chaudhuri, A. (2016). Empirical Bayes estimation method in some randomized response techniques. *Statistics and Applications, New Series, 15*(1,2), 101–116.

Singh, J. (1976). A note on randomized response techniques. In *Proceedings of the American Statistical Association (Social Statistics Section)* (Vol. 772).

Warner, S. L. (1965). Randomized response: A survey technique for eliminating evasive answer bias. *Journal of American Statistical Association, 60*, 63–69.

Winkler, R. L., & Franklin, L. A. (1979). Warner's randomized response model: A Bayesian approach. *Journal of the American Statistical Association, 74*(365), 207–214.

Chapter 6
Optional Randomized Response Technique (ORRT)

6.1 Introduction

Warner (1965) is the pioneer of the randomized response (RR) technique (RRT) which is used in the situation of sensitive and stigmatizing issues. There have been several extensions to this RR technique. An issue often raised with the RR technique is that some respondents are more willing to answer directly rather than their response to compulsory RR because the perception of sensitivity is not the same for every person. This is particularly true when studies are based on political party affiliation or avenues of savings. For this reason, a new RR technique, called optional randomized response (ORR) technique (ORRT) was developed.

An optional randomized response (ORR) technique (ORRT) permits sampled respondents either to directly divulge the stigmatizing characteristic or go for a randomized response respectively with complementary probabilities. The original concept of this design was described in Chaudhuri and Mukerjee (1985) with SRSWR. Chaudhuri and Dihidar (2009) extended this work to the general sampling design. Each sampled person here is approached with an option of (i) reporting directly whether he/she bears a sensitive characteristic and (ii) giving out an RR, by adopting the offered RR device. The direct response (DR) option '(i)' is executed with an unknown probability and the RR option '(ii)', with the complementary probability.

Chaudhuri and Mukerjee (1988), Arnab (2004), Chaudhuri and Saha (2005), Saha (2007) and many others suggested modifications to the ORR technique with a slightly different approach. Overall sample of respondents is classified into two parts to gather DR from one part and RR from another part. Corresponding to each part, linear unbiased estimators are derived which are combined further.

Gupta et al. (2002) proposed an ORRT permitting respondents to report true value or scramble response with a common probability, an unrealistic [as correctly pointed out first by Arnab (2004)] assumption. An estimator for the sensitivity level of the question is also derived there. Arnab (2004, 2018), Huang (2008, 2010) further extended the model discussing the procedures of Gupta et al. (2002). The following

A. Chaudhuri et al., *Randomized Response Techniques*,
https://doi.org/10.1007/978-981-99-9669-8_6

sections chronologically list the ORRTs for qualitative and quantitative stigmatizing characteristics.

6.2   Different Approaches—Equal and Varying Probabilities

Chaudhuri and Mukerjee (1985) suggested a plan of getting optional randomized responses rather than compulsory RR (CRR), giving each respondent the opportunity to give either a direct response (DR) to the question on stigmatizing characteristic or an RR. However, it was limited to SRSWR only. Gupta (2001), and Gupta et al. (2002) also studied ORR techniques in SRSWR design.

SRSWR is the simplest sampling design for finite population sampling. Any other alternative scheme of the sample-selection procedure is a complex sampling design. In real-life situations, complex sampling methods are put into practice. SRSWR is rather exceptional. The National Sample Survey Organization (NSSO) of India selects samples through varying probability sampling designs because the effectiveness of these designs is better than sampling with equal probabilities.

Chaudhuri and Saha's (2005) approach allowed drawing the sample from the population by a general sampling scheme. Chaudhuri and Dihidar (2009), Chaudhuri (2011a, 2011b) provided elaborative and enriching literature on ORRT.

6.2.1   Chaudhuri and Mukerjee (1985)

Chaudhuri and Mukerjee (1985) acknowledged the fact that the survey question may be sensitive for some respondents but may not be sensitive enough for other respondents. Therefore, a plan for optional randomized response (ORR) was suggested in the context of qualitative characteristics rather than compulsory randomized response (CRR) permitting an option to give either a direct response (DR) or randomized response (RR). However, the theories for RR and ORR were developed in multinomial set-up and for SRSWR sampling design.

Suppose a population can be classified into t disjoint categories C_j with unknown proportions $\pi_j(0 < \pi_j < 1)$, $j = 1, 2 \ldots t$. This classification is based on the character C such as affiliation to political parties, affection to athletic teams, etc. The interest is to estimate $\theta = G\pi$, where $\pi = (\pi_1, \pi_2, \ldots, \pi_t)$ and G is a $u \times t$ order known matrix.

Then for an RR technique, we may draw $m(\geq 1)$ independent SRSWR samples of sizes $n_1, n_2, \ldots, n_m$. Assume that for ith$(i = 1, 2, ..m)$ sample we have r_i distinct RR's and P_i is the known design matrix of order $r_i \times t$. Obviously, $1'_{r_i} P_i = 1'_t$.

Now, if no direct response (DR) is allowed, then for each respondent in the ith sample there are r_i possible responses and all are randomized. Then, $\mu_i = P_i \pi$ is the $r_i \times 1$ vector of probabilities of the different RR's with $\hat{\mu}_i$ as the corresponding vector of observed proportions $i = 1, 2, ..m$. The procedure admits a linear unbiased estimator of θ in terms of $\hat{\mu}_1, \hat{\mu}_2, \ldots, \hat{\mu}_m$ if and only if there exists a matrix B such that $G = BP$ where matrix $P = (P_1', P_2', \ldots, P_t')'$ of order $\left(\sum_i r\right) \times t$. This B can be partitioned as $B = (B_1, B_2, .., B_m)$, where B_i is $u \times r_i$ matrix and an unbiased estimator of θ is obtained as $\hat{\theta}^{(0)} = \sum_{i=1}^{m} B_i \hat{\mu}_i = B \hat{\mu}$.

This set-up further developed for ORR also, in Chaudhuri and Mukerjee (1985) and briefly described here considering dichotomous case under Warner model.

In ORR with Warner model, one may get four types of responses, say, (i) DR stating stigmatizing character A, (ii) DR stating A^c, (iii) RR stating 'yes' (iv) RR stating 'no'. Let in a random sample of size n, n_1, n_2, n_3, n_4 be the frequencies of responses of (i) to (iv).

Thus, the joint distribution of (n_1, n_2, n_3, n_4) is a multinomial having parameters n and $\lambda_1, \lambda_2, \lambda_3, \lambda_4$ with $\lambda_1 = \pi_1, \lambda_2 = \pi_2, \lambda_3 = p\pi_3 + (1 - p)\pi_4, \lambda_4 = (1 - p)\pi_3 + p\pi_4$.

Here, $\pi_1, \pi_2, \pi_3, \pi_4$ are the population proportions of people in the respective categories: (i) A and willing to disclose true value, (ii) A^c and willing to disclose true value, (iii) A and unwilling to disclose true value, (iv) A^c and unwilling to disclose true value.

Then, an unbiased estimator of $\pi = \pi_1 + \pi_3$, under ORR is $\tilde{\pi} = \frac{1}{n}\left[n_1 + \frac{pn_3 - (1-p)n_4}{2p-1}\right]$ and $V(\tilde{\pi}) = \frac{\pi(1-\pi)}{n} + \frac{p(1-p)}{n(2p-1)^2}(\pi_1 + \pi_3)$.

6.2.2 Chaudhuri and Mukerjee (1988)

Rao-Blackwellization was shown in Chaudhuri and Mukerjee's (1988) book to generate an improved way of estimation when certain respondents offered to divulge DRs when given the opportunity. The exact data collected from respondents is provided here for the DRs and separately for the RRs. Currently, a theory is being developed to give the respondent the choice of giving a DR rather than an RR without disclosing the actual option exercised to the investigator.

6.2.3 Gupta et al. (2002)

Gupta et al. (2002) and Huang (2010) considered ORRTs for quantitative characteristics in which the probability of selecting the randomized (or direct) response remains fixed for all respondents, say W and it is quite unacceptable. This W was referred to as a measure of sensitivity level of the question, and it is unknown. It can, however, be estimated approximately unbiasedly with a single sample, as briefly reported in Gupta et al. (2002). Furthermore, this approach can't be adopted for complex survey designs.

In Gupta et al. (2002), a sample of respondents is selected by SRSWR, and each of them is asked to choose one of the following options for the sensitive character Y:

1. to report true response Y
2. to provide scrambled response SY

with probability $(1 - W)$ and $W (0 \leq W \leq 1)$, respectively. Here, S is a scrambling variable, assumed to be a known distribution (like chi-square, exponential, binomial, Poisson) having mean $\mu_s = 1$ and variance $\sigma_s^2 = \gamma^2$. Also let μ_y be the mean of the sensitive character Y. This ORR model is a modification of the Eichhorn and Hayre (1983) model.

Thus, considering another random variable X as

$$X = \begin{cases} 1, & \text{if the response is scrambled} \\ 0, & \text{otherwise.} \end{cases}$$

The responses can be written as $Z = S^X Y$ and $E(X) = W$.

Now, $\mu_z = \frac{1}{n} \sum_{i=1}^{n} Z_i$ can be taken as an unbiased estimator of the mean of the sensitive character Y as

$$E(Z) = E\left(S^X Y\right) = E(SY)P(X = 1) + E(Y)P(X = 0)$$
$$= E(S)E(Y)P(X = 1) + E(Y)P(X = 0); \text{ since } E(S) = 1 \text{ and, } S \text{ and } Y \text{ are independent}$$
$$= \mu_y W + \mu_y(1 - W) = \mu_y.$$

The variance of the above estimator μ_z can be written as

$$V(\mu_z) = \frac{\sigma_z^2}{n} = \frac{1}{n}\left[\sigma_y^2 + W\gamma^2\left(\sigma_y^2 + \mu_y^2\right)\right],$$

and an estimator of $V(\mu_z)$ is

$$\widehat{V(\mu_z)} = \frac{1}{n}\left[s_y^2 + \hat{W}\gamma^2\left(s_y^2 + \mu_z^2\right)\right],$$

where $s_y^2 = \frac{s_z^2 - \hat{W}\gamma^2\mu_z^2}{1 + \hat{W}\gamma^2}$ and $s_z^2 = \frac{1}{n-1}\sum_{i \in s}(Z_i - \bar{z})^2$.

To the above formula, $\hat{W}$ is an estimator of W which can be computed as $\hat{W} = \frac{\frac{1}{n}\sum_{i=1}^{n} \log Z_i - \log \mu_z}{E(\log S)}$. However, it is a biased estimator because of the presence of logarithm terms. In addition $\widehat{V(\mu_z)}$ is merely a biased estimator.

Arnab (2004, 2018) extended Gupta et al. (2002) RR technique to overcome the above-mentioned weakness.

In nature, population is partitioned into two groups: non-sensitive $(\overline{G})$ and sensitive (G). Now, denoting Y_G and $Y_{\overline{G}}$ as the actual Y value of the respondents in group G and $\overline{G}$, respectively, it is quite obvious that $E(Y_G)$ and $E\left(Y_{\overline{G}}\right)$ are not same. Huang (2008) noticed such a potential difference between subpopulation which was ignored in Gupta et al. (2002) and suggested two simple modifications. The procedures are briefly described below.

1st Modification According to Huang (2008)

Two independent SRSWR samples of size n_i, $i = 1, 2$ are drawn from the population so that $n_1 + n_2 = n$. Each individual in the ith sample generates a random number $S_i (\neq 1)$ to report one of following options, decided by himself/ herself:

1. the true response Y
2. the scrambled response $S_i Y$.

Also let $\theta_i = 1$ and γ_i^2 be known mean and variance of S_i.
   Now, the expected value of sample response $Z_i (i = 1, 2)$ is

$$E(Z_i) = W E(S_i) E(Y_G) + (1 - W) E(Y_{\overline{G}}) = W \mu_G + (1 - W) \mu_{\overline{G}} = \mu,$$

where μ_G and $\mu_{\overline{G}}$ are the average vales of Y for groups G and $\overline{G}$, respectively, and μ is the population mean.

$$Similarly,\ E(Z_i^2) = W E(S_i^2) E(Y_G^2) + (1 - W) E(Y_{\overline{G}}^2)$$
$$= W(1 + \gamma_i^2)(\sigma_G^2 + \mu_G^2) + (1 - W)(\sigma_{\overline{G}}^2 + \mu_{\overline{G}}^2)$$
$$= W \gamma_i^2 (\sigma_G^2 + \mu_G^2) + \sigma^2 + \mu^2$$
$$owing\ to\ \sigma^2 + \mu^2 = W(\sigma_G^2 + \mu_G^2) + (1 - W)(\sigma_{\overline{G}}^2 + \mu_{\overline{G}}^2).$$

   Here, σ^2 is the population variance.
   Thus, $V(Z_i) = W \gamma_i^2 (\sigma_G^2 + \mu_G^2)$.
   If $\overline{Z}_1$ and $\overline{Z}_2$ are the means of two samples, Huang (2008) suggested $\hat{\mu}_I = \alpha \overline{Z}_1 + (1 - \alpha) \overline{Z}_2$ and $\hat{\sigma}_I^2 = \frac{\gamma_2^2 s_1^2 - \gamma_1^2 s_2^2}{\gamma_2^2 - \gamma_1^2}$ as unbiased estimators of μ and σ^2, respectively. Obviously, $\gamma_2^2 \neq \gamma_1^2$ and $0 \leq \alpha \leq 1$.
   In this paper, the following are also derived.

$$V(\hat{\mu}_I) = \alpha^2 \frac{\sigma_1^2}{n_1} + (1 - \alpha)^2 \frac{\sigma_2^2}{n_2},$$

$$\hat{V}(\hat{\mu}_I) = \alpha^2 \frac{s_1^2}{n_1} + (1 - \alpha)^2 \frac{s_2^2}{n_2}\ \ Where\ \ s_i^2 = \frac{1}{n_i - 1} \sum_{j=1}^{n_i} (Z_{ij} - \overline{Z}_i)^2$$

   An estimator of W is also given as $\hat{W}_I = \beta \hat{W}_1 + (1 - \beta) \hat{W}_2$ where $0 \leq \beta \leq 1$ and $\hat{W}_i = \frac{\sum_{j=1}^{n_i} \frac{\log Z_{ij}}{n_i} - \log \hat{\mu}_I}{E(\log S_i)}$.
   It should be noted that if $(\sigma_G^2 + \mu_G^2) = (\sigma_{\overline{G}}^2 + \mu_{\overline{G}}^2) = \sigma^2 + \mu^2$ and the choice of $\alpha = \beta = \frac{n_1}{n}$ results in same estimators of μ and W as in Gupta et al. (2002).

However, the suggested estimator of σ^2 remains better than Gupta et al.'s suggested estimator due to unbiasedness.

2nd Modification According to Huang (2008)

Here, the procedure is slightly different from the 1st modification, discussed above. The difference is that the known mean of S_i is $\theta_i \neq 1$ for $i = 1, 2$. The rest of the procedure is same as above.

Huang (2010) suggested a generalized ORR procedure with the unbiased estimations of mean, variance and sensitivity level. Gupta et al. (2010) is another extension of Gupta et al. (2002) to the two-stage ORR model combining the two-stage approach of Mangat and Singh (1990) and Ryu et al. (2006). Ryu et al. (2006) also studied two-stage optional randomized response assuming the W to be known.

6.2.4   Arnab (2004)

In this paper, Raghunath Arnab argued against Gupta et al.'s (2002) assumption of the same probability for selecting RR or DR for each individual and suggested an alternative ORR model considering the notion of Chaudhuri and Mukerjee (1988). Gupta et al.'s ORR model was extended further in Arnab (2004, 2018) to deal with any sampling design.

Here, a sample s with n individuals is selected by a sampling design P. If the ith$(i \in s)$ individual feels the character A is not confidential (i.e. $i \in$ non-sensitive group $\overline{G}$) will disclose the true value y_i. The respondent will provide an RR z_i if he/she feels A is confidential (i.e. $i \in$ sensitive group G).

Denoting r_i as the revised randomized response obtained from z_i only, such that $E_R(r_i) = y_i$, $V_R(r_i) = \sigma_i^2$ and $C_R(r_i, r_j) = 0 \forall i \neq j$, an unbiased estimator for the population mean $\mu_y = \frac{1}{N} \sum_{i=1}^{N} y_i$ was proposed there as follows:

$$t = \sum_{i \in s} b_{si} \tilde{r}_i,$$

taking $\tilde{r}_i = \delta_i y_i + (1 - \delta_i) r_i;.$ $\delta_i = \begin{cases} 1 & \text{if } i \in \overline{G} \\ 0 & \text{if } i \in G \end{cases}$ Here b_{si}'s are constants and independent of y_i satisfying the unbiasedness condition $\sum_{s \supset i} b_{si} p(s) = \frac{1}{N}$. This $p(s)$ denotes the probability of the selection of the sample s.

Arnab (2004) derived the variance of the estimator t as

$$V(t) = \sum_{i \in U} (\alpha_i - 1) y_i^2 + \sum_{i \neq} \sum_{j \in U} (\alpha_{ij} - 1) y_i y_j + \sum_{i \in G} \alpha_i \sigma_i^2,$$

where $\alpha_i = \sum_{s \supset i} b_{si}^2 p(s), \alpha_{ij} = \sum_{s \supset ij} b_{si} b_{sj} p(s)$.
An unbiased estimator of $V(t)$ is

$$\hat{V}(t) = \sum_{i \in s} d_{si} \tilde{r}_i^2 + \sum_{i \neq} \sum_{j \in s} d_{sij} \tilde{r}_i \tilde{r}_j + \sum_{s \cap G} d_{si}^* \hat{\sigma}_i^2 ,$$

where d_{si}, d_{sij} and d_{si}^* are free from $\tilde{r}_i$ and subject to $\sum_{s \supset i} d_{si} p(s) = \alpha_i - 1$; $\sum_{s \supset ij} d_{sij} p(s) = \alpha_{ij} - 1$; $\sum_{s \supset i} d_{si}^* p(s) = 1$.

For HT estimator, this $b_{si} = \frac{1}{N \pi_i}$.

Thus, the unbiased estimator $t = t_{HT} = \frac{1}{N} \sum_{i \in s} \frac{\tilde{r}_i}{\pi_i}$ and

$$V(t_{HT}) = \frac{1}{N^2} \left[\frac{1}{2} \sum_{i \neq} \sum_{j \in U} (\pi_i \pi_j - \pi_{ij}) \left(\frac{y_i}{\pi_i} - \frac{y_j}{\pi_j} \right)^2 + \sum_{i \in G} \frac{\sigma_i^2}{\pi_i} \right]$$

(Substituting $\alpha_i = \sum_{s \supset i} b_{si}^2 p(s) = \frac{1}{N^2 \pi_i}$ and $\alpha_{ij} = \sum_{s \supset ij} b_{si} b_{sj} p(s) = \frac{1}{N^2 \pi_{ij}}$)

$$\hat{V}(t_{HT}) = \frac{1}{N^2} \left[\frac{1}{2} \sum_{i \neq} \sum_{j \in s} \left(\frac{\pi_i \pi_j - \pi_{ij}}{\pi_{ij}} \right) \left(\frac{\tilde{r}_i}{\pi_i} - \frac{\tilde{r}_j}{\pi_j} \right)^2 + \sum_{i \in s \cap G} \frac{\hat{\sigma}_i^2}{\pi_i} \right]$$

***Extension of Gupta et al.** (2002) **RR According to Arnab** (2004)*

To extend Gupta et al. (2002) for complex survey design, let

$$z_i = \begin{cases} y_i & \text{with probability } (1 - W) \\ S_i y_i & \text{with probability } W \end{cases} ,$$

where S_i is a random variable with $E_R(S_i) = 1$ and $V_R(S_i) = \gamma^2$.

So, $E_R(z_i) = (1 - W) y_i + W E_R(S_i) y_i = y_i$. Thus, $r_i = z_i$.

Thus, $E_R(z_i^2) = E_R(r_i^2) = (1 - W) y_i^2 + W E_R(S_i^2) y_i^2 = (1 - W) y_i^2 + W(1 + \gamma^2) y_i^2$ and $V_R(r_i) = W \gamma^2 y_i^2 = \sigma_i^2$ and obviously, $C_R(r_i, r_j) = 0 \forall i \neq j$.

So, an unbiased estimator of μ_y for any sampling design can be written as

$$t^* = \sum_{i \in s} b_{si} r_i$$

with   the   variance $V(t^*)$ $=$ $\sum_{i \in U} (\alpha_i - 1) y_i^2$ + $\sum_{i \neq} \sum_{j \in U} (\alpha_{ij} - 1) y_i y_j + W \gamma^2 \sum_{i \in G} \alpha_i y_i^2$

Now, to estimate W, let

$$\log z_i = \begin{cases} \log y_i & \text{with probability } (1 - W) \\ \log(S_i y_i) & \text{with probability } W \end{cases}$$

and $E_R(\log z_i) = (1 - W) \log y_i + W E_R(\log S_i) + W \log y_i = \log y_i + W \delta$; taking $E_R(\log S_i) = \delta$ and it is expected to be known.

Also, defining $T^* = \sum_{i\in s} b_{si}\frac{\log z_i}{\delta}$ one gets $E(T^*) = E_P E_R(T^*) = \frac{1}{\delta N}\sum_{i\in U}\log y_i + W$.

Thus, $W = E(T^*) - \frac{1}{\delta N}\sum_{i\in U}\log y_i$.

Now, using the approximation $E(\log t^*) \approx \frac{1}{N}\sum_{i\in U}\log y_i$ in view of Gupta et al. (2002), Arnab (2004) suggested an estimator of W as

$$\hat{W} = T^* - \frac{1}{\delta}\log t^* = \frac{\sum_{i\in s} b_{si}\log z_i - \log(\sum_{i\in s} b_{si} z_i)}{\delta}$$

For SRSWR, the above expression becomes $\hat{W} = \frac{\frac{1}{n}\sum_{i\in s}\log z_i - \log(\frac{1}{n}\sum_{i\in s} z_i)}{\delta}$ which was obtained by Gupta et al. (2002).

Extension of Gupta et al. (2002) RR According to Arnab (2018)

Under this RR technique, assuming x_i as a random sample from a known distribution with mean μ_x and variance σ_x^2, let

$$z_i = \begin{cases} y_i, & i \in \overline{G} \\ Q_i y_i, & i \in G \end{cases} \text{ where } Q_i = \frac{x_i}{\mu_x}$$

Therefore, $z_i = \delta_i y_i Q_i + (1-\delta_i)y_i$, where $\delta_i = \begin{cases} 1 & i \in \overline{G} \\ 0 & i \in G \end{cases}$, following Arnab

(2004). Clearly, $E_R(z_i) = y_i$ and $V_R(z_i) = \delta_i y_i^2 V_R(Q_i) = \delta_i y_i^2\frac{\sigma_x^2}{\mu_x^2}$.

Then,

$$t_{HT}^{*G} = \frac{1}{N}\sum_{i\in s}\frac{z_i}{\pi_i}$$

is the Horvitz-Thompson estimator of population mean μ_y having variance

$$V\left(t_{HT}^{*G}\right) = \frac{1}{N^2}\left[\frac{1}{2}\sum_{i\neq}\sum_{j\in U}(\pi_i\pi_j - \pi_{ij})(\frac{y_i}{\pi_i} - \frac{y_j}{\pi_j})^2 + \frac{\sigma_x^2}{\mu_x^2}\sum_{i\in G}\frac{y_i^2}{\pi_i}\right],$$

and unbiased estimator of variance is

$$\hat{V}\left(t_{HT}^{*G}\right) = \frac{1}{N^2}\left[\frac{1}{2}\sum_{i\neq}\sum_{j\in s}\left(\frac{\pi_i\pi_j - \pi_{ij}}{\pi_{ij}}\right)(\frac{z_i}{\pi_i} - \frac{z_j}{\pi_j})^2 + \frac{\frac{\sigma_x^2}{\mu_x^2}}{1 + \frac{\sigma_x^2}{\mu_x^2}}\sum_{i\in s\cap G}\frac{z_i^2}{\pi_i}\right].$$

6.2.5 Chaudhuri and Saha (2005)

The main objective of this paper was to develop an unbiased estimator of finite population proportion covering a sensitive character while a sample is chosen with varying probability sampling design and the responses are taken through ORRT.

Let a sample s be chosen from a finite population $U = (1, 2 \ldots . N)$. Suppose s_1 is the subsample of s in which the people don't feel the survey question is sensitive enough and ready to divulge their true values. Therefore, knowing these values an estimator can be developed for $i \in s_1$ which is $\sum_{i \in s_1} y_i b_{si} I_{si}$. Another estimator can also be developed for $i \in s - s_1$ using their RR, say r_i and it is $\sum_{i \in s-s_1} r_i b_{si} I_{si}$.

Thus, $t = \sum_{i \in s_1} y_i b_{si} I_{si} + \sum_{i \in s-s_1} r_i b_{si} I_{si}$ is an unbiased estimator for population total, $Y = \sum_{i=1}^{N} y_i$. The variance of the estimator and an unbiased estimator of the variance were derived there.

6.2.6 Pal (2008)

In this paper, an optional randomized response model was developed using Eichhorn and Hayre (1983) scrambled response model gathering two independent responses from each individual. An unbiased estimator of population total ($\sum_{i=1}^{N} y_i$) with its variance and unbiased estimator of variance were derived for varying probability sampling scheme.

The respondents, selected by a sampling design, are requested to report their first response as the true value with an unknown probability or the scrambled response with the complementary probability. Instructions to report second independent response are also given that time. To record the second independent response, it is requested to each respondent to report one of the following:

1. the true value y_i or
2. randomized response using the second RR device and this RR device also has two options. The respondent is either to report the earlier scrambled response with a known probability say $p(0 < p < 1)$ that was already recorded as first response or to report a new scrambled response with probability $(1 - p)$.

The approach of Pal (2008) is quite different from others as she only permits the respondents to record their first scrambled response as a second independent response with a known probability.

Let Z_i and Z_i' be the two independent responses for the ith person in the sample s. Thus,

$$
Z_i = \begin{cases}
y_i, & \text{with the unknown probability } c_i \, (0 \le c_i \le 1) \\
I_i = \frac{y_i x_i}{\theta_1} + u_i, & \text{with the complementary probability } (1 - c_i)
\end{cases} \quad i = 1, 2, \ldots, n,
$$

and

$$Z_i' = \begin{cases} y_i, & \text{with the unknown probability} c_i \\ I_i & \text{with the probability} (1 - c_i)p \\ I_i' = \frac{y_i x_i'}{\theta_2} + u_i' & \text{with the probability} (1 - c_i)(1 - p) \end{cases}$$

Here, $x(x')$ is a random variable independent of y having known mean $\theta_1(\theta_2)(\neq 0)$ and variance $\sigma_1^2(\sigma_2^2)$. Let $u(u')$ be another random variable independent of $x(x')$ with known mean $\alpha_1(\alpha_2)$ and variance $\gamma_1^2(\gamma_2^2)$. Also let $x_i(x_i')$ and $u_i(u_i')$ be the random values of the random variables $x(x')$ and $u(u')$.

Now,

$$E_R(Z_i) = y_i c_i + (1 - c_i)E_R(I_i) = y_i c_i + (1 - c_i)\left(\frac{y_i \theta_1}{\theta_1} + \alpha_1\right) = y_i + (1 - c_i)\alpha_1,$$

and

$$\begin{aligned} E_R\left(Z_i'\right) &= y_i c_i + (1 - c_i)pE_R(I_i) + (1 - c_i)(1 - p)E_R\left(I_i'\right) \\ &= y_i + (1 - c_i)(p(\alpha_1 - \alpha_2) + \alpha_2) \\ &= y_i + (1 - c_i)\alpha_2' \quad (\text{Substituting } p(\alpha_1 - \alpha_2) + \alpha_2 = \alpha_2') \end{aligned}$$

Taking $r_i = \frac{\alpha_2' Z_i - \alpha_1 Z_i'}{\alpha_2' - \alpha_1}$ it can be easily shown that $E_R(r_i) = y_i$ and $V_R(r_i) = V_i = E_R\left(r_i^2\right) - y_i^2$.

So, an unbiased estimator of $\sum_{i=1}^{N} y_i$ is $e = \sum_{i \in s} r_i b_{si}$ having variance

$$V(e) = E_P V_R(e) + V_P E_R(e) = E_P\left(\sum_{i \in s} V_i b_{si}^2\right) + V_P\left(\sum_{i \in U} y_i b_{si}\right),$$

and unbiased estimator of $V(e)$ as $v(e) = \sum_{i \in s} v_i b_{si} + v_p(e)$, briefly described in Pal (2008).

Not only that, this paper also prescribed an estimator of c_i as $\hat{c}_i = 1 - \frac{Z_i - Z_i'}{\alpha_1 - \alpha_2'}$ which is unbiased.

6.2.7   Chaudhuri and Dihidar (2009)

Chaudhuri and Dihidar (2009) proposed an ORR under general sampling design where each respondent generates two independent responses each either direct or randomized according to his/her choice. Qualitative and quantitative characteristics, both the cases are considered here for finite population.

Let $U = (1, 2, \ldots N)$ be a finite population and y_i be the unknown value of the study variable y. Also let for an ORR device, $c_i (0 < c_i < 1 \forall i \in U)$ be an unknown probability of the ith person who chooses to report directly, without revealing his/her option to the enquirer.

Further, let the response for ith person be

$$I_i = \begin{cases} DR, & \text{with probability } c_i \\ RR \text{ with a specific device,} & \text{with probability } (1 - c_i) \end{cases},$$

and

$$I_i' = \begin{cases} DR, & \text{with probability } c_i \\ RR \text{ with another specific device,} & \text{with probability } (1 - c_i) \end{cases}$$

The investigator is to explain to each sampled person a formal way to implement choosing such an undisclosed $c_i (0 < c_i < 1 \forall i \in U)$ and $(1 - c_i)$ to be the probability of reporting a direct response (DR) and respectively a randomized response (RR) with no option to change it for an alternative RR device. An estimator of y_i, say r_i is taken as a linear combination of those two independent responses and obviously,

$$E_R(r_i) = y_i,$$

$$E(r_i) = E_P E_R(r_i) = E_R E_P(r_i),$$

and

$$V(r_i) = E_P V_R(r_i) + V_P E_R(r_i) = E_R V_P(r_i) + V_R E_P(r_i)$$

Here, E_P, V_P are taken as operators for expectation and variance due to the sampling design P. E_R and V_R are the operators for expectation and variance due to the randomized response.

This paper presented the above version of optional randomized response technique for Warner (1965) RR device, Greenberg et al. (1969) and for Mangat and Singh (1990).

6.3 Qualitative Characteristics

In this section, we have illustrated the ORR technique of the above version for Warner (1965), Greenberg et al. (1969), Boruch (1971) and Kuk (1990) RR devices.

6.3.1 ORRT Using Warner's RRT

Each sampled person is instructed to report his/her true value of the stigmatizing attribute A directly or by Warner's RR device. The RR device is defined as follows. A box containing a number of identical cards marked A or A^c in the proportions $p_1:1 - p_1(0 < p_1 < 1, p_1 \neq \frac{1}{2})$ is given to the sampled person with a request to draw a card from the box and to say truthfully whether the selected card type matches or not the person's characteristic, without divulging the card type drawn. The procedure is repeated one more time independently but with a different Warner's RR device with another similar box. Here, the cards are marked by A or A^c in the proportions $p_2:1 - p_2(0 < p_2 < 1, p_2 \neq \frac{1}{2}, p_1 \neq p_2)$.

Thus, two independent optional randomized responses for ith$(i \in U)$ person are Z_i and Z_i'.

Here,

$$z_i = \begin{cases} y_i, & \text{with unknown probability } c_i \\ \text{Warner's RR,} & \text{with unknown probability } (1 - c_i), \text{ using the first box.} \end{cases}$$

and

$$z_i = \begin{cases} y_i, & \text{with unknown probability } c_i \\ \text{Warner's RR with unknown probability} (1 - c_i), \text{ using another box} \end{cases}.$$

Then, writing RR-based expectations and variances as E_R and V_R,

$$E_R(Z_i) = c_i y_i + (1 - c_i)[p_1 y_i + (1 - p_1)(1 - y_i)]$$

$$E_R(Z_i) = c_i y_i + (1 - c_i)[p_1 y_i + (1 - p_1)(1 - y_i)]$$
$$E_R(Z_i') = c_i y_i + (1 - c_i)[p_2 y_i + (1 - p_2)(1 - y_i)].$$

Thus, an unbiased estimator of y_i is

$$r_i = \frac{(1 - p_2)Z_i - (1 - p_1)Z_i'}{p_1 - p_2}, \quad p_1 \neq p_2,$$

and an unbiased estimator for the variance $V_R(r_i)$ is
$v_i = \frac{(1-p_1)(1-p_2)}{(p_1-p_2)^2}(Z_i - Z_i')^2$. The details of the proof are given below.

Variance Estimation of the ORR Model Using Warner's RRT

As noted earlier, the estimator of y_i is $r_i = \frac{(1-p_2)Z_i - (1-p_1)Z_i'}{p_1-p_2}$, variance $V_R(r_i)$ of the estimator r_i is given by

$$V_R(r_i) = \frac{(1 - p_2)^2 V_R(Z_i) + (1 - p_1)^2 V_R(Z_i')}{(p_1 - p_2)^2},$$

where

$$V_R(Z_i) = E_R\left(Z_i^2\right) - [E_R(Z_i)]^2 = E_R(Z_i)[1 - E_R(Z_i)] \quad \left[\text{putting } Z_i^2 = Z_i\right]$$
$$= [c_i y_i + (1 - c_i)\{(1 - p_1) + (2p_1 - 1)y_i\}]$$
$$\left[1 - c_i y_i - (1 - c_i)\{(1 - p_1) + (2p_1 - 1)y_i\}\right]$$
$$= c_i y_i + (1 - c_i)(1 - p_1) + (1 - c_i)(2p_1 - 1)y_i - c_i^2 y_i - c_i(1 - c_i)$$
$$(1 - p_1)y_i - c_i(1-c_i)(2p_1 - 1)y_i - c_i(1 - c_i)(1 - p_1)y_i - (1 - c_i)^2(1 - p_1)^2 -$$
$$- (1 - c_i)^2(1 - p_1)(2p_1 1)y_i - (1 - c_i)^2(2p_1 - 1)^2 y_i \quad \left[\text{putting } y_i^2 = y_i\right]$$
$$= (1 - c_i)(1 - p_1)[c_i + (1 - c_i)p_1].$$

Similarly, $V_R\left(Z_i'\right) = (1 - c_i)(1 - p_2)[c_i + (1 - c_i)p_2]$.
So,

$V_R(r_i)$

$$= \frac{(1 - p_2)^2(1 - c_i)(1 - p_1)(c_i + (1 - c_i)p_1) + (1 - p_1)^2(1 - c_i)(1 - p_2)(c_i + (1 - c_i)p_2)}{(p_1 - p_2)^2}$$
$$= \frac{(1 - p_1)(1 - p_2)}{(p_1 - p_2)^2}(1 - c_i)[(1 - p_2)(c_i + (1 - c_i)p_1) + (1 - p_1)(c_i + (1 - c_i)p_2)]$$
$$= \frac{(1 - p_1)(1 - p_2)}{(p_1 - p_2)^2}(1 - c_i)[(2 - p_1 - p_2) - 2(1 - c_i)(1 - p_1)(1 - p_2)].$$

Now,

$$E_R(Z_i - Z_i')^2 = E_R(Z_i) + E_R\left(Z_i'\right) - 2E_R(Z_i)E_R\left(Z_i'\right) \quad \left[\text{putting } Z_i^2 = Z_i\right]$$
$$= (1 - c_i)[(2 - p_1 - p_2) - 2(1 - c_i)(1 - p_1)(1 - p_2)].$$

Thus,

$$\frac{(1 - p_1)(1 - p_2)}{(p_1 - p_2)^2}E_R\left(Z_i - Z_i'\right)^2 = V_R(r_i),$$

i.e. $v_i = \frac{(1-p_1)(1-p_2)}{(p_1-p_2)^2}\left(Z_i - Z_i'\right)^2$ is an unbiased estimator for $V_R(r_i)$.
The above-mentioned form of unbiased variance estimator is slightly different from Chaudhuri and Dihidar (2009). According to them, $V_R(r_i) = E_R\left(r_i^2\right) - E_R(r_i) = E_R(r_i(r_i - 1))$. So, $v_i^* = r_i(r_i - 1)$ is an unbiased estimator of $V_R(r_i)$, which is obvious. They also claimed that this estimator v_i^* may be negative. However, with a little algebra, it can be shown that $v_i = v_i^*$.

It is a function of two independent responses (Z_i, Z_i'). The possible values of (Z_i, Z_i') are (1, 1) (0, 0) (1, 0) and (0, 1). It should be noted that the different responses 1 or 0 coming from the same person for the first and second trials reveal that the person has opted for an RR. But it does not reveal the person's actual status of the stigmatizing character.

6.3.2 ORRT Using Greenberg et al.'s RRT

Here, the ORR technique with this RR device is described as follows.

Two boxes, box 1 and box 2, containing a number of cards marked A, the stigmatizing characteristic or B, the innocuous characteristic in different known proportions $p_1 : (1 - p_1)$ and $p_2 : (1 - p_2)$ are given to sampled persons. Each sampled person is instructed to draw one card from each box independently if he/she opts for RR and report according to the above devices.

So, the optional randomized response for ith person is

$$Z_i = \begin{cases} y_i, & \text{with unknown probability } c_i \\ \text{Greenberg et al. RR,} & \text{with unknown probability } (1 - c_i), \text{ using box 1} \end{cases}$$

and

$$Z_i' = \begin{cases} y_i, & \text{with unknown probability } c_i \\ \text{Greenberg et al. RR,} & \text{with unknown probability } (1 - c_i), \text{ using box 2.} \end{cases}$$

Defining

$$x_i = \begin{cases} 1, & \text{if } i\text{th person bears } B \\ 0, & \text{if } i\text{th person bears } B^c \end{cases}.$$

It can be written that

$$P(Z_i = 1) = c_i y_i + (1 - c_i)[p_1 y_i + (1 - p_1)x_i],$$

and

$$P(Z_i = 0) = c_i(1 - y_i) + (1 - c_i)[p_1(1 - y_i) + (1 - p_1)(1 - x_i)].$$

Hence, taking RR-based expectation it is obtained that

$$E_R(Z_i) = c_i y_i + (1 - c_i)[p_1 y_i + (1 - p_1)x_i].$$

Similarly,

$$E_R(Z_i') = c_i y_i + (1 - c_i)[p_2 y_i + (1 - p_2)x_i].$$

Thus, an unbiased estimator of y_i under the ORR technique with Greenberg et al.'s (1969) RRT is

$$r_i = \frac{(1 - p_2)Z_i - (1 - p_1)Z_i'}{p_1 - p_2} \text{taking } p_1 \neq p_2,$$

and the variance $V_R(r_i)$ is given by

$$V_R(r_i) = \frac{(1 - p_2)^2 V_R(Z_i) + (1 - p_1)^2 V_R(Z_i')}{(p_1 - p_2)^2},$$

where $V_R(Z_i) = E_R(Z_i)(1 - E_R(Z_i)) = (y_i - x_i)^2(1 - p_1)(1 - c_i)(p_1 + (1 - p_1)c_i)$
and $V_R(Z_i') = (y_i - x_i)^2(1 - p_2)(1 - c_i)(p_2 + (1 - p_2)c_i)$.

Therefore, $V_R(r_i) = \frac{(y_i - x_i)^2(1 - c_i)(1 - p_1)(1 - p_2)}{(p_1 - p_2)^2}[2c_i(1 - p_1)(1 - p_2) + p_1 + p_2 - 2p_1 p_2]$.

Thus, an unbiased estimator for $V_R(r_i)$ is

$$v_i = \frac{(1 - p_1)(1 - p_2)}{(p_1 - p_2)^2}(Z_i - Z_i')^2 \text{ since } p_1 \neq p_2$$

as $E_R(Z_i - Z_i')^2 = (y_i - x_i)^2(1 - c_i)[2c_i(1 - p_1)(1 - p_2) + p_1 + p_2 - 2p_1 p_2]$.

6.3.3 ORRT Using Boruch's RRT

In the ORR technique with Boruch's (1971) Forced RR model, ith sampled person is directed to report the direct response y_i with unknown probability $c_i(0 \leq c_i \leq 1)$ or the forced RR with the complementary probability $1 - c_i$. Each sampled person is offered two boxes—box 1 and box 2, with a number of cards marked as "yes", "no" and "Honest Response" but in different proportions $p_1 : p_2:(1 - p_1 - p_2)$ and $p_3 : p_4:(1 - p_3 - p_4)$, respectively. Here, $0 < p_1, p_2, p_3, p_4 < 1$ of course. Also, $p_1 + p_2 < 1$ and $p_3 + p_4 < 1$ are the compulsory criteria for defining proportions. Another restriction $p_1 p_4 = p_2 p_3$ is needed on the known probabilities p_1, p_2, p_3 and p_4, to derive an unbiased estimator.

So, for ith person, the optional randomized responses are

$$Z_i = \begin{cases} y_i, & \text{with unknown probability } c_i \in [0, 1] \\ \text{Forced RR,} & \text{with unknown probability } (1 - c_i), \text{ using box 1} \end{cases},$$

and

$$Z_i' = \begin{cases} y_i, & \text{with unknown probability } c_i \in [0, 1] \\ \text{Forced RR,} & \text{with unknown probability } (1 - c_i), \text{ using box 2.} \end{cases}$$

Then,

$$P(Z_i = 1) = c_i y_i + (1 - c_i)[(1 - p_1 - p_2)y_i + p_1],$$

and

$$P(Z_i = 0) = c_i(1 - y_i) + (1 - c_i)[(1 - p_1 - p_2)(1 - y_i) + p_2].$$

Similarly,

$$P(Z'_i = 1) = c_i y_i + (1 - c_i)[(1 - p_3 - p_4)y_i + p_3],$$

and $P(Z'_i = 0) = c_i(1 - y_i) + (1 - c_i)[(1 - p_3 - p_4)(1 - y_i) + p_4]$.

Now, an unbiased estimator for y_i is

$$r_i = \frac{p_3 Z_i - p_1 Z'_i}{p_3 - p_1}$$

having variance $V_R(r_i) = \frac{p_3^2}{(p_3-p_1)^2} V_R(Z_i) + \frac{p_1^2}{(p_3-p_1)^2} V_R(Z'_i)$, where $V_R(Z_i) = (2y_i - 1)(1 - c_i)\{(p_1 + p_2)y_i - p_1\} - (1 - c_i)^2\{(p_1 + p_2)y_i - p_1\}^2$ and $V_R(Z'_i) = (2y_i - 1)(1 - c_i)\{(p_3 + p_4)y_i - p_3\} - (1 - c_i)^2\{(p_3 + p_4)y_i - p_3\}^2$.

Now,

$$E_R(Z_i - Z'_i)^2$$

$$= p_1 p_3 (2y_i - 1)(1 - c_i)\left(\frac{p_1 + p_2}{p_1}y_i - 1\right)\left(\frac{1}{p_1} + \frac{1}{p_3} - 2(1 - c_i)\right)$$

$$\left(\frac{p_1 + p_2}{p_1}y_i - 1\right)(2y_i - 1) = \frac{(p_3 - p_1)^2}{p_1 p_3} V_R(r_i).$$

Hence, an unbiased estimator for $V_R(r_i)$ can be written as

$$v_i = \frac{p_1 p_3 (Z_i - Z'_i)^2}{(p_3 - p_1)^2}.$$

6.3.4   ORRT Using Kuk's RRT

Pal et al. (2020) developed ORRT with Kuk model. Here, each respondent is requested to report the true value of bearing the stigmatizing character A using the ORR technique adopting Kuk's RR device. The RR devices are defined as follows:

The respondent is instructed to draw k (with replacement) number of cards from one of the two boxes—box 1 and box 2, containing red and black cards in different proportions $\theta_1 : (1 - \theta_1)$ and $\theta_2 : (1 - \theta_2)$, respectively. It is obvious that $0 < \theta_1, \theta_2 < 1$. The respondent is therefore requested to report the number $\frac{\left(\frac{f_i}{k} - \theta_2\right)}{\theta_1 - \theta_2}$, where f_i is the number of red cards out of the selected k cards, if the person opts for Kuk's RR device. These k cards should be drawn from box 1, if the person bears the stigmatizing character, otherwise, the cards are drawn from box 2 without divulging which box is used to draw the cards.

So, for ith person, the response is

$$
Z_i = \begin{cases} y_i, & \text{with unknown probability } c_i \in [0, 1] \\ \dfrac{\left(\frac{f_i}{k} - \theta_2\right)}{\theta_1 - \theta_2}, & \text{with the unknown probability } (1 - c_i). \end{cases}
$$

Clearly, $E_R(f_i) = k[y_i\theta_1 + (1 - y_i)\theta_2]$ leads to

$$
E_R(Z_i) = c_i y_i + (1 - c_i) E_R \left(\frac{\frac{f_i}{k} - \theta_2}{\theta_1 - \theta_2} \right) = y_i.
$$

This ORR procedure is needed to be repeated one more time to estimate the variance. Suppose, the second response variable is Z_i' which is the same as above, but the number of red cards drawn at the second time is denoted by f_i'.

So, the final unbiased estimator of y_i becomes

$$
r_i = \frac{Z_i + Z_i'}{2}.
$$

Following Chaudhuri et al. (2013), an unbiased variance estimator is

$$
v_i = \frac{1}{4}(Z_i - Z_i')^2.
$$

Pal et al. (2020) considers a data consisting of a fictitious set of 116 undergraduate students aged below 20 and their reckless driving history with weekly expenditures to assess the performance of the proposed ORR models with various RR devices.

Treating y as a qualitative variable related to a stigmatizing characteristic—"Fined for breaking traffic rules", the aim of interest is to estimate the proportion of the students who broke the traffic rules, last year. Let the population proportion is known, and the value is 0.8275. In the dataset, an auxiliary variate x—"Interested in painting", is taken for the numerical illustration of the ORR technique with Greenberg et al. (1969) model. This variate is unrelated to the variable of interest. Size-measure variable z—"Weekly expenditure", is used here to draw samples in varying probability sampling design.

To show the competitiveness concerning the proposed ORR models, samples of size $n = 39$ are taken from the finite population U by Lahiri (1951), Midzuno (1952), Sen (1953) sampling scheme where the first unit is selected with the probability $p_i^* = \frac{z_i}{\sum_U z_i}$ and the remaining units are selected by Simple Random Sampling (SRS) Without Replacement (SRSWOR) from the remaining units in U, after the first draw. Now, employing the HT (1952) method of estimation, the estimator of population proportion (π_A) may be defined as

$$e_{HT} = \frac{1}{N} \sum_{i=1}^{n} \frac{y_i}{\pi_i}.$$

Here, the inclusion probability $\pi_i = p_i^* + \left(1 - p_i^*\right)\frac{n-1}{N-1}$.

Since y_i is not assessable directly due to its stigmatizing characteristic, an unbiased estimator r_i of y_i is assigned here.

Hence, the final unbiased estimator of π_A is defined as

$$e_{HT,RR} = \frac{1}{N} \sum_{i=1}^{n} \frac{r_i}{\pi_i}$$

and an unbiased estimator of variance is given by

$$v\left(e_{HT,RR}\right) = \frac{1}{N^2}\left[\sum_{i<j\in s}\sum \frac{\pi_i \pi_j - \pi_{ij}}{\pi_{ij}}(\frac{r_i}{\pi_i} - \frac{r_j}{\pi_j})^2 + \sum_{i\in s} \frac{v_i}{\pi_i} \right],$$

where v_i is an unbiased estimator of the variance of r_i and the second-order inclusion probability $\pi_{ij} = \frac{(n-1)(N-n)\left(p_i^*+p_j^*\right)+(n-1)(n-2)}{(N-1)(N-2)}$

This $v\left(e_{HT,RR}\right)$ is always positive in Lahiri-Midzuno-Sen Sampling scheme.

For each of the proposed models, 1000 samples are drawn from U using the mentioned sampling scheme and, after performing the proposed ORR methods, estimates are obtained.

Efficacies of the proposed ORR models are judged by the Average Coverage Probabilities (ACP), the Average Coefficient of Variation (ACV) and the Average Length (AL) of the 95% confidence interval (CI) based on $e_{HT,RR} \pm 1.96\sqrt{v(e_{HT,RR})}$. The point estimator will be judged well if the ACV—the average over 1000 replications of estimated coefficient of variations $\left(CV = 100 \times \frac{\sqrt{v\left(e_{HT,RR}\right)}}{e_{HT,RR}} \right)$, has a small magnitude. The value less than 10% or at most 30% is preferable. The percentage of cases for which 95% CI covers the true value of the parameter is called 'ACP'. ACP values (in %) close to 95% will be preferred. 'AL' is defined as $2 \times 1.96\sqrt{v(e_{HT,RR})}$. Smaller ACV, AL values along with the ACP value close to 95% will be preferred.

Results based on the above-mentioned criteria are given in Tables 6.1, 6.2, 6.3 and 6.4 for four different ORRTs.

6.4 Quantitative Characteristics

Several authors have contributed to the theory of quantitative characteristics based ORR techniques. Gupta et al. (2002) and Pal (2008), respectively, extended Eichhorn and Hayre's (1983) method in ORRT using SRSWR and varying probability

Table 6.1   ACV, ACP and AL for ORRT with Warner's model

p_1	p_2	ACV	ACP	AL
0.28	0.19	48.1865	94.6	3.8623
0.49	0.36	37.5586	99.1	2.0169
0.54	0.29	26.3733	98	1.0650
0.63	0.56	42.3268	97.3	2.6278
0.66	0.45	24.7465	97.9	0.9712
0.77	0.65	26.8792	99.4	1.0953
0.81	0.63	20.2835	92.5	0.7123

Table 6.2   ACV, ACP and AL for ORRT with Greenberg et al.'s model

p_1	p_2	ACV	ACP	AL
0.36	0.23	42.2166	95.2	2.4704
0.51	0.39	38.2907	86.8	2.0682
0.56	0.49	45.1583	86.5	3.0992
0.69	0.54	28.0311	98.3	1.1740
0.72	0.55	24.9143	93.6	0.9746
0.88	0.34	11.7061	96.8	0.3454
0.91	0.61	12.1766	93.5	0.3634

Table 6.3   ACV, ACP and AL for ORRT with Boruch's model

p_1	p_2	p_3	p_4	ACV	ACP	AL
0.64	0.23	0.24	0.09	33.7349	90.1	0.5066
0.45	0.4	0.52	0.46	46.4117	79.4	3.1305
0.39	0.31	0.43	0.34	55.3157	76.4	4.6584
0.25	0.4	0.37	0.59	27.8805	91.2	1.1824
0.32	0.38	0.4	0.48	37.776	84.4	2.0412
0.35	0.23	0.47	0.31	32.5996	87.7	1.5525
0.15	0.13	0.22	0.19	28.2077	98.1	1.1993

Table 6.4   ACV, ACP and AL for ORRT with Kuk's model

θ_1	θ_2	ACV	ACP	AL
0.6	0.2	5.4740	94.5	0.1749
0.8	0.6	11.3493	95.4	0.3933
0.56	0.4	12.1329	96	0.3856

sampling designs. This section is entirely devoted to the advancements of several ORR techniques for quantitative characteristics associated with social stigmas, using varying probability sampling designs. Considering the approach of Chaudhuri and Dihidar (2009), the implementation procedures of those techniques are given here. The estimation procedures are also presented in the respective sections.

Some numerical illustrations are presented at the end of this section to study the effectiveness and competitiveness of the proposed quantitative ORR models.

6.4.1 ORR Using Eichhorn and Hayre's RR Device

Pal (2008) worked on the ORR technique with the Scramble Response model by capturing two independent responses for each respondent. For the first response, the ORR technique permitted a sampled respondent either to directly divulge the stigmatizing characteristic with an unknowable probability or to give a scramble response with its complementary probabilities. However, for the second RR device, it was suggested here either to report a new scrambled response or the RR value obtained through the first RR device. Saha (2011) proposed some modifications of the ORR technique with the Scramble Response model for stratified varying probability sampling design.

Here, the proposed ORR method is a modification of Pal (2008) and the work published in Patra and Pal (2019). Let y_i be the true sensitive value of ith respondent ($i = 1, 2, ..., N$), and x be the discrete random variable with known mean θ_1 and variance σ_1^2. Also, let ψ be another discrete random variable which is independent to x, having known mean 0 and variance σ_2^2. Considering the fact that some respondents may wish to answer directly to the sensitive question with unknowable probability c_i, a choice of DR (y_i) and a RR (I_i) is given to them, instead of the compulsory RR. The procedure is known as the ORR technique which is discussed earlier in Chap. 2.

In this ORR method, $I_i = \frac{y_i x_i}{\theta_1} + \psi_i$ while x_i and ψ_i are the values of the random variables x and ψ respectively for ith respondent.

Mathematically, the response of ith person may be written as:

$$Z_i = \begin{cases} y_i \text{ with unknown probability } c_i \in [0, 1] \\ \frac{y_i x_i}{\theta_1} + \psi_i \text{ with unknown probability } 1 - c_i. \end{cases}$$

Now, denoting E_R as RR based expectation and V_R as RR based variance, it is found that

$$E_R(Z_i) = c_i y_i + (1 - c_i) \left\{ \frac{y_i \theta_1}{\theta_1} + 0 \right\} = y_i$$

And

$$V_R(Z_i) = E_R(Z_i^2) - E_R(Z_i)^2$$

$$= c_i y_i^2 + (1 - c_i) \left\{ \frac{y_i^2}{\theta_1^2}(\sigma_1^2 + \theta_1^2) + \sigma_2^2 + 2\frac{y_i\theta_1}{\theta_1}.0 \right\} - y_i^2$$

$$= (1 - c_i) \left(\frac{y_i^2\sigma_1^2}{\theta_1^2} + \sigma_2^2 \right)$$

Obviously, the unbiased estimator of y_i is Z_i and the variance $V_R(Z_i) = (1 - c_i)\left(\frac{y_i^2\sigma_1^2}{\theta_1^2} + \sigma_2^2\right)$ is unknown as y_i is not assessable directly.

Thus, in order to estimate the variance, the whole process is repeated one more time independently to get another response Z_i'. The interpenetrating network of subsampling technique led by P. C. Mahalanobis (1946) is used here to provide the final RR based estimator of y_i, which becomes

$$r_i = \frac{Z_i + Z_i'}{2}$$

and a variance estimator is written as

$$v_i^* = \frac{1}{4}(Z_i - Z_i')^2.$$

6.4.2 ORR Using Chaudhuri's RR Device I

Here, each sampled person is instructed to report his/her true response to the question relating to quantitative sensitive characteristics either directly or by Chaudhuri's RR *Device I*. For an ORR device, the ith person may secretly exercise the direct response option with unknown probability $c_i (0 \le c_i \le 1 \forall i \in U)$ or may give out the randomized response with probability $(1 - c_i)$. This process is repeated independently twice, with similar RR devices having different RR parameters. The RR devices are defined as follows.

An RR device is constructed here with two boxes, Box 1 and Box 2. Box 1 contains $T(> 1)$ number of identically designed cards bearing real numbers $a_1, a_2, a_3, ...a_T$ together with mean $\mu_a = \frac{1}{T}\sum_i a_i = 1$. Box 2 contains $M(> 1)$ number of identically designed cards with real numbers $b_1, b_2, b_3, ..., b_M$. The sampled person is instructed to draw one card from each box if the person opts for RR and to say $Z_i = a_j y_i + b_k; (j = 1, 2 ..., Tk = 1, 2, ..M)$ without divulging the number a_j and b_k, visible on the selected cards from box 1 and box 2, respectively. Another RR device is also constituted with two boxes, namely box 3 and box 4. Box 3 contains T' number of identically designed cards bearing real numbers $a_1', a_2', ..., a_{T''}'$ such that $\mu_{a'} = \frac{1}{T'}\sum_{i=1}^{T'} a_i' = 1$. Box 4 contains M' number of identically designed cards bearing real numbers $b_1', b_2', ... b_{M''}'$ but the mean $\mu_{b'} = \frac{1}{M'}\sum_{i=1}^{M'} b_i'$ should not be equal to $\mu_b = \frac{1}{M} \sum_{i=1}^{M} b_i$. The sampled person is instructed to draw one card from each box present in this RR device and to report the number $Z'_i = a'_j y_i + b'_k$; $(j = 1, 2 \ldots, T'k = 1, 2, \ldots, M')$ without revealing the numbers a'_j and b'_k visible on the selected cards from the boxes.

Therefore, the ORRs for ith person are

$$Z_i = \begin{cases} y_i & \text{with unknown probability} c_i \in [0, 1] \\ a_j y_i + b_k & \text{with unknown probability} (1 - c_i) \end{cases},$$

and

$$Z'_i = \begin{cases} y_i & \text{with unknown probability} c_i \in [0, 1] \\ a'_j y_i + b'_k & \text{with unknown probability} (1 - c_i). \end{cases}$$

Taking RR-based expectation, it is obtained that

$$E_R(Z_i) = c_i y_i + (1 - c_i)(\mu_a y_i + \mu_b)$$

and

$$E_R(Z'_i) = c_i y_i + (1 - c_i)(\mu_{a'} y_i + \mu_{b'}).$$

Thus, an unbiased estimator of y_i is

$$r_{1i} = \frac{\mu_{b'} Z_i - \mu_b Z'_i}{\mu_{b'} - \mu_b}; \mu_{b'} \neq \mu_b.$$

To get an estimator of the variance, the whole process should be repeated independently one more time and the responses for ith person are denoted by (G_i, G'_i). Therefore, another unbiased estimator of y_i is

$$r_{2i} = \frac{\mu_{b'} G_i - \mu_b G'_i}{\mu_{b'} - \mu_b}; \mu_{b'} \neq \mu_b.$$

Now, for ith person, final RR-based estimator of y_i is given by

$$r_i^* = \frac{r_{1i} + r_{2i}}{2}$$

such that $E_R(r_i^*) = y_i$ and the variance estimator is written as

$$v_i^* = \frac{1}{4}(r_{1i} - r_{2i})^2.$$

6.4.3 ORR Using Chaudhuri's RR Device II

In the case of the ORR technique considering Chaudhuri's RR *device II*, each sampled person is approached with a box of several identically designed cards. A choice of DR or RR is given to them. The sampled person may opt for the DR with unknowable probability $c_i \in [0, 1]$, if he/she is willing to answer the direct query.

There are k (known) proportion of cards in the box, marked as "true". The remaining cards bear a real number from $x_1, x_2, ..., x_M$ which are present in the box with known proportions $q_1, q_2, ..., q_M$, respectively, such that $\sum_{i=1}^{M} q_i = 1 - k$. Each sampled person is requested to draw a card randomly and report the true value if the person gets a card with a visible mark "true", or otherwise report the number x_j ($j = 1, 2, ..., M$) visible on the selected card.

So, the ORR for ith person is

$$Z_i = \begin{cases} y_i & \text{with unknown probability } c_i \\ y_i & \text{with unknown probability } (1 - c_i)k \\ x_j & \text{with unknown probability } (1 - c_i)q_j \end{cases}.$$

Taking RR-based expectation, it is obtained that

$$E_R(Z_i) = c_i y_i + (1 - c_i)\left(ky_i + \sum_{j=1}^{M} q_j x_j \right).$$

To get an unbiased estimator of y_i, the process is repeated independently one more time with a different set of cards in different proportions and the response is recorded here as z_i'. Clearly,

$$E\left(Z_i'\right) = c_i y_i + (1 - c_i)\left(k'y_i + \sum_{j=1}^{M'} q_j' x_j' \right)$$

Using Eqs. (3.5) and (3.6) with the multiplication of $\left(1 - k'\right)$ and $(1 - k)$, respectively, it is found that

$$E_R\left((1 - k')Z_i - (1 - k)Z_i'\right)$$
$$= \left(k - k'\right)y_i + (1 - c_i)\left[(1 - k')\sum_{j=1}^{M} q_j x_j - (1 - k)\sum_{j=1}^{M'} q_j' x_j' \right]$$

Therefore, an unbiased estimator of y_i can be derived if $(1 - k')\sum_{j=1}^{M} q_j x_j - (1 - k)\sum_{j=1}^{M'} q_j' x_j' = 0$.

That implies

$$E_R\left[\frac{(1-k')Z_i - (1-k)Z_i'}{(k-k')}\right] = y_i; \, if \, \frac{1-k'}{1-k} = \frac{\sum_{j=1}^{M'} q_j' x_j'}{\sum_{j=1}^{M} q_j x_j}.$$

Thus, an unbiased estimator of y_i is.

$r_{1i}' = \frac{(1-k')Z_i - (1-k)Z_i'}{(k-k')}$. In order to estimate the variance, the whole process should be repeated independently one more time and then another response is denoted by (G_i, G_i'). The interpenetrating network of subsampling technique is used here to provide the final RR-based estimator of y_i, which is written as

$$r_i = \frac{r_{1i}' + r_{2i}'}{2},$$

and a variance estimator is given by

$$v_i^* = \frac{1}{4}\left(r_{1i}' - r_{2i}'\right)^2.$$

Here, the data containing the reckless driving history of a fictitious set of $N = 116$ undergraduate students aged below 20 and their weekly expenditures is used to judge the efficacy of the proposed quantitative ORR models with Eichhorn and Hayre (1983) RR device and Chaudhuri's RR device I.

The amount that had been fined in last year is considered here as the quantitative sensitive variable, $say y$. Our aim of interest is to estimate the average amount of fine (π_A^*), they paid last year. Here, the population mean is $\pi_A^* = $ Rs.908.069/- .

Samples of size $n = 39$ are taken from U using Lahiri (1951), Midzuno (1952), Sen (1953) varying probability sampling design. An auxiliary variable z, 'Weekly expenditure', is used here to draw samples in varying probability sampling design which is known as the size-measure variable. Since y represents a stigmatizing characteristic, the true value of y_i for $ith \forall i \in s$ person is not directly assessable. It may be estimated through the ORR device.

Considering r_i to be an unbiased estimator for y_i, Horvitz-Thompson (1952) estimator

$$e = \frac{1}{N}\sum_{i=1}^{n}\frac{r_i}{\pi_i}$$

is employed in estimating the average amount of fine π_A^*.

Thus, an unbiased estimator of variance becomes

$$v(e) = \frac{1}{N^2}\left[\sum_{i<j\in s}\sum \frac{\pi_i\pi_j - \pi_{ij}}{\pi_{ij}}\left(\frac{r_i}{\pi_i} - \frac{r_j}{\pi_j}\right)^2 + \sum_{i\in s}\frac{v_i^*}{\pi_i}\right]$$

Table 6.5 ACV, ACP and AL for ORR with Eichhorn and Hayre RR device

θ_1	ACV	ACP	AL
1	11.27102	99.8	424.3873
1.85	10.06902	95.4	311.3131
2.5	8.16192	89.7	253.4241
4	8.12986	92	331.1016
6	8.61532	94.4	326.0284
10	10.9889	96	416.0089

Table 6.6 ACV, ACP and AL for ORR with Chaudhuri's RR device I

μ_b	$\mu_{b'}$	ACV	ACP	AL
5	6	16.9961	100	437.7337
5.5	4	9.6232	94	466.2567
3	6	8.5731	96.3	328.5278
3.5	4.5	8.1771	95.7	573.2008
6.4	3.8	8.3414	96.9	224.686
6.7	7.5	11.4735	95.6	404.1155

A large number of samples, say $T = 1000$, are drawn from the population U to judge the efficacy of the proposed ORR models through the Average Coverage Probability (ACP), the Average Coefficient of Variation (ACV), and the Average Length (AL) of the 95% confidence intervals.

Following Tables 6.5 and 6.6 represents the ACV (in %), ACP (in %) AL for the ORR models with Eichhorn and Hayre (1983) RR device and Chaudhuri (2011a, 2011b) RR device I.

References

Arnab, R. (2004). Optional randomized response techniques for complex designs. *Biometrical Journal, 46*, 114–124.

Arnab, R. (2018). Optional randomized response techniques for quantitative characteristics. *Communications in Statistics- Theory and Methods*.

Boruch, R. F. (1971). Assuring confidentiality of responses in social research: A note on strategies. *American Sociologist, 6*, 308–311.

Chaudhuri, A. (2011a). *Randomized response and indirect questioning techniques in surveys*. CRC Press.

Chaudhuri, A. (2011a). Unequal probability sampling: Analyzing optional randomized response on qualitative and quantitative variables bearing social stigma. In *International Statistical Institute Proceedings of the 58th World Statistics Congress, 2011*, (pp. 1939–1947). Dublin.

Chaudhuri, A., & Dihidar, K. (2009). Estimating means of stigmatizing qualitative and quantitative variables from discretionary responses randomized or direct. *Sankhya B, 71*, 123–136.

Chaudhuri, A., & Mukerjee, R. (1985). Optionally randomized responses techniques. *Calcutta Statistical Association Bulletin, 34*, 225–229.

Chaudhuri, A., & Mukerjee, R. (1988). *Randomized responses: Theory and techniques*. Marcel Dekker.

Chaudhuri, A., & Saha, A. (2005). Optional versus compulsory randomized response techniques in complex surveys. *Journal of Statistical Planning and Inference, 135*, 516–527.

Eichhorn, B., & Hayre, L. S. (1983). Scrambled randomized response methods for obtaining sensitive quantitative data. *Journal of Statistical Planning and Inference, 7*, 307–316.

Greenberg, B. G., Abul-Ela, A. L., Simmons, W. R., & Horvitz, D. G. (1969). The unrelated question randomized response model: Theoretical framework. *Journal of American Statistical Association, 64*, 520–539.

Gupta, S. (2001). Qualifying the sensitivity level of binary response personal interview survey questions. *Journal of Combinatorics, Information and System Sciences, 26*, 101–109.

Gupta, S., Gupta, B., & Singh, S. (2002). Estimation of sensitive level of personal interview survey questions. *Journal of Statistical Planning and Inference, 100*, 239–247.

Gupta, S., Shabbir, J., & Sehra, S. (2010). Mean and sensitivity estimation in optional randomized response models. *Journal of Statistical Pranning and Inference, 140*(10), 2870–2874.

Horvitz, D. G., & Thompson, D. J. (1952). A generalization of sampling without replacement from a finite universe. *Journal of American Statistical Association, 47*, 663–685.

Huang, K. (2008). Estimation for sensitive characteristics using optional randomized response technique. *Quality and Quantity, 42*, 679–686.

Huang, K. C. (2010). Unbiased estimators of mean, variance and sensitivity level for quantitative characteristics in finite population sampling. *Metrika, 71*, 341–352.

Kuk, A. Y. (1990). Asking sensitive questions indirectly. *Biometrika, 77*(2), 436–438.

Lahiri, D. B. (1951). A method of sample selection providing unbiased ratio estimates. *Bulletin of International Statistical Institute, 3*(2), 133–140.

Mahalanobis, P. C. (1946). Recent experiments in statistical sampling in the Indian Statistical Institute. *Journal of the Royal Statistical Society, 109*, 325–378.

Mangat, N. S., & Singh, R. (1990). An alternative randomized response procedure. *Biometrika, 77*, 439–442.

Midzuno, H. (1952). On the sampling system with probability proportional to the sum of the sizes. *Annals of Institute of Statistical Mathematics, 3*, 99–107.

Pal, S. (2008). Unbiasedly estimating the total of a stigmatizing variable from a complex survey on permitting options for direct or randomized responses. *Statistical Papers, 49*, 157–164.

Pal, S., Chaudhuri, A., & Patra, D. (2020). How privacy may be protected in optional randomized response surveys. *Statistics in Transition, 21*(2), 61–87.

Patra, D., & Pal, S. (2019). Privacy protection in optional randomized response surveys for quantitative characteristics. *Staistics and Applications, 17*(2), 47–54.

Ryu, J., Kim, J., Heo, T., & Park, C. (2006). On stratified randomized response sampling. *Model Assisted Statistics and Applications, 1*, 31–36.

Saha, A. (2007). Optional randomized response in stratified unequal probability sampling: A simulation based numerical study with Kuk's method. *TEST, 16*, 346–354.

Sen, A. R. (1953). On the estimator of the variance in sampling with varying probabilities. *Journal of Indian Society of Agricultural Statistics, 5*, 119–127.

Warner, S. L. (1965). Randomized response: A survey technique for eliminating evasive answer bias. *Journal of American Statistical Association, 60*, 63–69.

Chapter 7
Protection of Privacy

7.1 Introduction

An RR or ORR survey's goal is to increase survey participation with respondents' actual values, which will lead to the production of good estimates for sensitive characteristics from the statisticians' perspective. As the respondents' true state of nature is covered by the RR device, it is important to understand whether the procedure assures that respondents' true nature will not be surely identified, i.e. a protection of privacy. Of course, participation will increase with enhanced protection. However, it should be emphasized that no universally acceptable measure is available yet. Privacy measures for RRT in the case of the SRSWR method of selection have been studied by many researchers. These include Lanke (1975, 1976), Leysieffer and Warner (1976), Anderson (1975a, 1975b, 1975c), Diana and Perii (2013). Chaudhuri and Saha (2004) followed by Chaudhuri et al. (2009), covered the general sampling procedures. In case of the quantitative sensitive attribute, Bose (2015) presented a privacy measure. Recently, Gupta et al. (2018) prescribed a unified measure of quantitative RR model quality with regard to the model comparison, inspired by Yan et al. (2008). Bose and Dihidar (2018) suggested privacy measures based on Kulbak-Leibler and Hellinger distance measures. Sections 7.2 and 7.3 illustrate the measures for qualitative RRTs.

7.2 Protection of Privacy Measures for Qualitative RRTs

Assume A is a sensitive attribute and $P(A)$ represents the probability that a person selected at random from the population U bears the attribute A. Let $P(A|R)$ $(or\, P(A^c|R))$ be the conditional probability of the sensitive attribute A $(or\, A^c)$ given the response R.

7.2.1 Leysieffer and Warner (1976)'s Protection Measure

Leysieffer and Warner (1976) suggested a jeopardy measure by quantifying the probability of an observation belonging to the sensitive trait A and its complement A^c while giving his/ her response as R, termed as revealing probabilities. Thus, $g(R, A) = \frac{P(A|R)/P(A)}{P(A^c|R)/1-P(A)} = \frac{P(R|A)}{P(R|A^c)}$ may be defined as the measure of jeopardy of R with respect to A. Similar to this, let $g(R, A^c) = \frac{P(R|A^c)}{P(R|A)}$ be the jeopardy measure of R with respect to A^c. When the value of $g(R, A)$ is more (less) than unity, the response R is in jeopardy to $A(A^c)$. This function g is now known as the jeopardy function for the given procedure.

7.2.2 Lanke (1976)'s Protection Measure

Lanke (1976) stated that respondents would seek to hide their true nature in A but not in A^c if A only the stigmatizing characteristic and considered the conditional probability $P(A|R)$ to quantify a different type of privacy measure. According to this, the quantity $g(A) = \max\{P(A|R = yes), P(A|R = no)\}$ may be utilized for the comparison of two methods if the respondents are chosen using SRSWR. From his perspective, the lower value of $g(A)$ is believed to be more protective than others. His study provides us with a suitable Warner's parameter given Simmons's parameter, allowing us to infer that both RR techniques are equally protective.

Fligner et al. (1977) also suggested three measures of respondent protection and described the strategies to incorporate these into a comparison of the RR techniques. Defining $(1 - \max[P_s(A|yes), P_s(A|no)])/(1 - P(A))$ as 'primary protection' measure by a RR technique S, they compared Warner's RRT and URL. By requiring only the primary protection of these two RRTs' to be equal, the criterion for comparison was the mean square error of both estimates of population proportion.

7.2.3 Anderson (1975a)'s Privacy Measure

Anderson (1975a) considered the conditional probabilities $P(A|R)$ and $P(A^C|R)$ as "risk of suspicion" and suggested the following restrictions to the conditional probabilities $P(A|R)$ and $P(A^c|R)$,

$$P(A|R) \leq \varepsilon_2 < 1 \quad \text{and} \quad P\left(A^c|R\right) \leq 1 - \varepsilon_1 < 1,$$

i.e. $\varepsilon_1 \leq P(A|R) \leq \varepsilon_2$.

That implies $g(R = yes, A) \leq \frac{1-P(A)}{P(A)} \frac{\varepsilon_2}{1-\varepsilon_2}$ and $g(R = no, A^c) \leq \frac{P(A)}{1-P(A)} \frac{1-\varepsilon_1}{\varepsilon_1}$.

7.2.4 Nayak (1994)'s Privacy Measure (Qualitative)

Considering $P(yes|A) = a$, $P(no|A^c) = b$ and prior probability $P(A) = \theta$, Nayak (1994) observed that the posterior probabilities for assessing respondent's protection can be expressed as $P(A|yes) = \frac{\theta a}{\theta a + (1-\theta)(1-b)}$ and $P(A|no) = \frac{\theta(1-a)}{\theta(1-a)+(1-\theta)b}$.

Therefore, the departure of $P(A|yes)$ from θ and $P(A^c|no)$ from $1-\theta$ is relevant to measure privacy. In Nayak (1994), a jeopardy measure is considered in the case of the sample-selection procedure by SRSWR.

7.2.5 Chaudhuri et al. (2009)'s Privacy Measure (Qualitative)

With the approach in Chaudhuri (2001), Chaudhuri et al. (2009) extended Nayak (1994)'s jeopardy measure as a response-specific measure of jeopardy, $J_i(R)$, which is depending upon the specific response R of the ith respondent. Then,

$$J_i(R) = \frac{L_i(R)\big/ L_i}{(1 - L_i(R))\big/ (1 - L_i)},$$

and

$$\overline{J_i} = \frac{1}{|R|} \sum_R J_i(R),$$

where L_i is the unknowable prior probability that the ith respondent bears the sensitive attribute A and $L_i(R)$ denotes the posterior probability that given the RR denoted R, the respondent bears A. $|R|$ is the number of possible responses of the specific RR model. It should be noted that $\overline{J_i}$ does not depend on L_i and the $\overline{J_i}$ close to unity means the better the protection of privacy. To examine the situations, some well-known qualitative RR models like Warner (1965), Greenberg et al. (1969), Kuk (1990), etc., are considered.

Warner's RR Device

The posterior probabilities at the given responses $R = 0$ and $R = 1$ are as follows:

$$\begin{aligned}
L_i(0) &= \frac{L_i P(I_i = 0 | y_i = 1)}{L_i P(I_i = 0 | y_i = 1) + (1 - L_i) P(I_i = 0 | y_i = 0)} \\
&= \frac{(1-p)L_i}{(1-p)L_i + p(1-L_i)} = \frac{(1-p)L_i}{p + (1 - 2p)L_i},
\end{aligned}$$

and

$$L_i(1) = \frac{L_i P(I_i = 1 | y_i = 1)}{L_i P(I_i = 1 | y_i = 1) + (1 - L_i)P(I_i = 1 | y_i = 0)} = \frac{pL_i}{(1 - p) + (2p - 1)L_i}.$$

As $p \to \frac{1}{2}$, $L_i(0) \to L_i$ and $L_i(1) \to L_i$ are desirable.
Therefore,

$$J_i(0) = \frac{L_i(0)/L_i}{(1 - L_i(0))/(1 - L_i)} = \frac{1 - p}{p} \quad \text{and} \quad J_i(1) = \frac{L_i(1)/L_i}{(1 - L_i(1))/(1 - L_i)} = \frac{p}{1 - p}.$$

In addition, $\overline{J}_i = \frac{1}{2}\left(\frac{1-p}{p} + \frac{p}{1-p}\right)$.

If $p = \frac{1}{2}$, $J_i(0) = J_i(1) = \overline{J}_i = 1$ which is desirable for an RR device. But $p = \frac{1}{2}$ is not permissible.

Greenberg et al.'s RR Device

Possible responses are $(0, 0)$, $(1, 0)$, $(0, 1)$ and $(1, 1)$.

The posterior probabilities are

$L_i(0, 0)$

$$= \frac{L_i P(I_i = 0 | y_i = 1) P(I_i' = 0 | y_i = 1)}{L_i P(I_i = 0 | y_i = 1) P(I_i' = 0 | y_i = 1) + (1 - L_i)L_i P(I_i = 0 | y_i = 0) P(I_i' = 0 | y_i = 0)}$$

$$= \frac{L_i(1 - p_1)(1 - p_2)}{L_i(1 - p_1)(1 - p_2) + (1 - L_i)p_1 p_2},$$

$L_i(0, 1)$

$$= \frac{L_i P(I_i = 0 | y_i = 1) P(I_i' = 1 | y_i = 1)}{L_i P(I_i = 0 | y_i = 1) P(I_i' = 1 | y_i = 1) + (1 - L_i)L_i P(I_i = 0 | y_i = 0) P(I_i' = 1 | y_i = 0)}$$

Similarly, $L_i(1, 0) = \frac{p_1(1 - p_2)L_i}{L_i p_1(1 - p_2) + (1 - L_i)p_2(1 - p_1)}$,

and $L_i(1, 1) = \frac{p_1 p_2 L_i}{p_1 p_2 L_i + (1 - L_i)(1 - p_1)(1 - p_2)}$.

As, $p_1 + p_2 \to 1$, $L_i(1, 1)$ and $L_i(0, 0) \to L_i$. But all the posterior probabilities tend to L_i if $p_1, p_2 \to \frac{1}{2}$.

Now, the response-specific jeopardy measure can be written here as follows:

$$J_i(R, R') = \frac{L_i(R, R')/L_i}{(1 - L_i(R, R'))/(1 - L_i)} \forall (R, R')$$

$$\overline{J}_i = \frac{1}{4}[J_i(0, 0) + J_i(0, 1) + J_i(1, 0) + J_i(1, 1)]$$

$$= \frac{1}{4}\left[\frac{(1 - p_1)(1 - p_2)}{p_1 p_2} + \frac{p_2(1 - p_1)}{p_1(1 - p_2)} + \frac{p_1(1 - p_2)}{p_2(1 - p_1)} + \frac{p_1 p_2}{(1 - p_1)(1 - p_2)}\right].$$

This $\overline{J}_i \to \frac{1}{2} + \frac{1}{4}\left[\left(\frac{p_1}{1-p_1}\right)^2 + \left(\frac{1-p_1}{p_1}\right)^2\right]$ if $p_1 + p_2 \to 1$. But, if $p_1, p_2 \to \frac{1}{2}$, all $J_i(R, R') \to 1$.

Forced Response Device

$$L_i(0) = \frac{L_i p_2}{L_i p_2 + (1 - L_i)(1 - p_1 - p_2 + p_2)} = \frac{L_i p_2}{L_i p_2 + (1 - L_i)(1 - p_1)},$$

and $L_i(1) = \dfrac{L_i(1 - p_1 - p_2 + p_1)}{L_i(1 - p_1 - p_2 + p_1) + (1 - L_i)p_1} = \dfrac{L_i(1 - p_2)}{L_i(1 - p_2) + (1 - L_i)p_1}.$

If $p_1 + p_2 \to 1$, $L_i(0)$ and $L_i(1) \to L_i$.

Therefore, $J_i(0) = \frac{p_2}{1-p_1}$ and $J_i(1) = \frac{1-p_2}{p_1}$ tend to unity with the condition $p_1 + p_2 \to 1$.

Kuk's RR Device

If the ith respondent is instructed to draw k cards (with replacement) from the box 1 or box 2 and report f_i as the number of red cards out of the selected k cards, the posterior probability and response-specific jeopardy measure can be written as follows:

$$
\begin{aligned}
L_i(f_i) &= \frac{L_i P(I_i = f_i | y_i = 1)}{L_i P(I_i = f_i | y_i = 1) + (1 - L_i)P(I_i = f_i | y_i = 0)} \\
&= \frac{L_i \theta_1^{f_i}(1 - \theta_1)^{k-f_i}}{L_i \theta_1^{f_i}(1 - \theta_1)^{k-f_i} + (1 - L_i)\theta_2^{f_i}(1 - \theta_2)^{k-f_i}}
\end{aligned}
$$

and

$$J_i(f_i) = \frac{\theta_1^{f_i}(1 - \theta_1)^{k-f_i}}{\theta_2^{f_i}(1 - \theta_2)^{k-f_i}}.$$

In addition, $\overline{J}_i = \frac{1}{k+1}\sum_{f_i=0}^{k} J_i(f_i)$.

If $\theta_1 \to \theta_2$, $L_i(f_i) \to L_i$ and $J_i(f_i) \to 1 \; \forall f_i$.

7.3 Protection of Privacy Measures for Quantitative RRTs

7.3.1 Anderson (1977)'s Privacy Measure

Anderson (1977) presented a measure of privacy protection using the distribution function of the quantitative attribute (γ), say $F_\gamma(t)$. As the randomized response $R = r$ depends on the unknowable value of γ, the response density can be represented as h $(r|\gamma = t)$ for a given value $\gamma = t$.

Thus, the unconditional density of R can be written as $g(r) = \int h(r|\gamma = t)dF_\gamma(t)$.

Then, the conditional density of t given the response $R = r$ is $f_\gamma(t|R) = \frac{h(r|\gamma=t)f_\gamma(t)}{g(r)}$.

Here, $f_\gamma(t)$ is the marginal density of γ, and $f_\gamma(t|R)$ represents the revealing density. Higher concentration of $f_\gamma(t|R)$ about the true value γ indicates that the respondent's privacy is not protected well there. This paper suggested to consider $V(\gamma|r)$ or $\frac{V(\gamma|r)}{V(\gamma)}$ as a measure of privacy protection with the overall measure $E\{V(\gamma|r)\}$ or $\frac{E\{V(\gamma|r)\}}{V(\gamma)}$, respectively.

7.3.2 Eichhorn and Hayre (1983)'s Privacy Measure

With the approach of the scramble response model, Eichhorn and Hayre (1983) proposed a privacy measure to evaluate how effectively the model works. Their suggestion is based on the ratio of the upper limit and lower limit of $100(1 - \alpha)\%$ confidence interval for the mean of the scrambling variable. For a given α, the larger the ratio the greater the protection.

7.3.3 Yan et al. (2008)'s Privacy Measure

As per their approach, one may allow $\Delta = E(r - y)^2$ as a measure of protection of the respondents' privacy. Here, r denotes the respondent's response through the quantitative RR model, whereas y is the actual value of our query.

7.3.4 Diana et al. (2013)'s Privacy Measure

In this paper, they discussed several quantitative RR models and some well-known privacy measures in a unified manner. They have considered a randomized linear model $X = V_1Y + V_2$ in which different choices of the two random variables V_1 and

V_2 represent several quantitative RR models like Pollock and Bek (1976), Eichhorn and Hayre (1983), etc. They have noted the concept of differential entropy where Shannon's information theory has been used. Also, the measures of privacy based on the correlation coefficient (Diana & Perii 2008; Zhimin et al., 2010) and multiple correlation coefficients (Diana & Perii, 2011) have been cited in this writing with others. Those privacy measures are expressed there in terms of the randomized linear model.

7.3.5 Chaudhuri and Christofides (2013)'s Privacy Measure

Evaluation of the respondents' privacy protection in case of quantitative sensitive issues has been studied in Chaudhuri and Christofides (2013)'s monograph. This Chapter of this monograph covers a brief review of the protection of privacy measures from the very beginning. Following the arguments in Chaudhuri (2001) and Chaudhuri et al. (2009), the Bayes' theorem is applied here. Then, the posterior probability of y_i for the given value of z_i turns out to be

$$L(y_i|z_i) = \frac{L_i P(z_i|y_i)}{P(z_i)},$$

where $P(z_i|y_i)$ denotes the conditional probability of reported quantitative RR value (z_i) for the ith person while the true response is y_i.

The degree of privacy protection measure is maximum if the value of the measure approaches L_i.

7.3.6 Bose (2015)'s Privacy Measure (Discrete Sensitive Variable)

Bose (2015) suggested the protection of privacy measures exclusively on SRSWR in the case of the discrete quantitative variable to estimate the population mean. To do so, she considered the sensitive variable of interest, X assuming the values $x_1, x_2, ..., x_m$ and R the randomized response variable such that the ranges of X and R are the same. If $P(X = x_k)$ denotes the probability of true response x_k of a sampled person and $P(X = x_k|R = x_j)$ is the conditional probability of true response $X = x_k$ while given a randomized response $R = x_j$, termed as revealing probability. Then, the protection of privacy measure is defined as follows:

$$\alpha = \max_{1 \leq k, j \leq m} \left| P(X = x_k|R = x_j) - P(X = x_k) \right|.$$

Thus, the lower value α provides a higher level of protection in privacy. This measure is also mentioned in Chaudhuri and Christofides (2013)'s monograph.

7.3.7 Bose and Dihidar (2018)'s Privacy Measure (Continuous Sensitive Variable)

This literature focused on the respondents' privacy protection for some RR techniques accessible for continuous sensitive variables. The distance from "true" and "revealing distributions" was used to generate measures for Pollock and Bek's (1976) additive model and Eichhorn and Hayre's (1983) multiplicative model. This article also highlighted the method of RR device parameters consideration to ensure a stipulated level of privacy protection.

Consider Pollock and Bek's (1976) additive model where the response R is taken as the sum of the continuous sensitive variable X and scramble variable S from known distribution, i.e. $R = X + S$. Also assume X follows normal distribution and S is taken as normal distribution with known mean μ_S and variance σ_S^2. Then, the joint distribution of X and R follows bivariate normal distribution with parameters $\mu_X, \mu_X + \mu_S, \sigma_X^2, \sigma_X^2 + \sigma_S^2, \sigma_X^2$.

Let f be the density function of X, called the 'true' density and g be the conditional density of $X|R = r$, called the 'revealing' density. Then, the comparison of these densities indicates how much a respondent's privacy is compromised. For this purpose, Kulback-Leibler divergence measure is a well-accepted method. Here, the measure can be written as

$$k = \int_{-\infty}^{\infty} \left\{ \log_e \left(\frac{f}{g} \right) \right\} f \, dx = \log_e \frac{\sigma}{\sigma_X} - \frac{1}{2} + \frac{1}{2\sigma^2} \{ \sigma_X^2$$

$$+ \frac{\sigma_X^4}{(\sigma_X^2 + \sigma_S^2)^2} (r - \mu_X - \mu_S)^2 \};$$

$$\sigma^2 = \frac{\sigma_X^2 \sigma_S^2}{\sigma_X^2 + \sigma_S^2}.$$

Thus, $E_R(k) = \log_e \frac{\sigma}{\sigma_X} - \frac{1}{2} + \frac{1}{2\sigma^2} \left\{ \sigma_X^2 + \frac{\sigma_X^4}{(\sigma_X^2 + \sigma_S^2)^2} \right\} = \frac{\sigma_X^2}{\sigma_S^2} - \log_e \sqrt{1 + \frac{\sigma_X^2}{\sigma_S^2}}$ measures the closeness between the true and revealing densities on an average.

Now, $1 - \left(\frac{\sigma_S^2}{\sigma_X^2} \right) E_R(k) \to 0$ as $\sigma_S^2 \to 0$ and $1 - \left(\frac{\sigma_S^2}{\sigma_X^2} \right) E_R(k)$ approaches 0.5 if $\sigma_S^2 \to \infty$.

Using the above points, Bose and Dihidar (2018) proposed a measure of privacy protection as

$$D_1 = 2 \times \frac{\sigma_S^2}{\sigma_X^2} \times log_e \sqrt{1 + \frac{\sigma_X^2}{\sigma_S^2}},$$

and this D_1 lies between 0 and 1. Thus, the closer D_1 is to 1 the higher the protection of respondents' privacy.

Another measure D_2 is also proposed here based on the square of Hellinger distance measure,

$$H^2(f, g) = 1 - \int_{-\infty}^{\infty} \sqrt{f(x)g(x)} = 1 - \sqrt{\frac{2\sigma_X\sigma}{\sigma_X^2 + \sigma^2}} \exp\left\{ \frac{1}{4} \frac{\sigma_X^4 (r - \mu_X - \mu_S)^2}{(\sigma_X^2 + \sigma_S^2)^2 (\sigma_X^2 + \sigma^2)} \right\}.$$

Now, $E_R\left(H^2(f, g)\right) = 1 - 2\sqrt{\frac{\sigma\left(\sigma_X^2 + \sigma_S^2\right)}{\sigma_X\left(3\sigma_X^2 + 4\sigma_S^2\right)}}$ and the measure for privacy protection

is $D_2 = 1 - E_R\left(H^2(f, g)\right) = 2\sqrt{\frac{\sigma\left(\sigma_X^2 + \sigma_S^2\right)}{\sigma_X\left(3\sigma_X^2 + 4\sigma_S^2\right)}}$. Clearly, $0 \le D_2 \le 1$, and the value

of D_2 close to 1 indicates the higher privacy protection.

These two measures (D_1 and D_2) can be derived for Eichhorn and Hayre (1983)'s multiplicative model, and it is also illustrated in Bose and Dihidar (2018).

7.4 Protection of Privacy Measures for Qualitative ORRT

7.4.1 Pal et al. (2020)'s Privacy Measure

It should be noted that the above-mentioned mathematical measures are not alone sufficient for studying the protection of privacy of ORRTs. As only one response employing the ORR technique is inadequate, two independent responses are needed when Chaudhuri and Mukerjee (1985)'s ORR technique is to be employed. Pal et al. (2020) developed the privacy measure for ORRT, while each sampled person is requested to give two ORRs independently with different known RR devices.

Denoting the responses as R and R', the posterior probability and the measure of jeopardy for ith person may be written as $L_i(R, R')$ and $J_i(R, R')$, respectively.

Now, applying Bayes' Theorem,

$$Prob\left(A|(R, R')\right) = \frac{L_i Prob(R|A)Prob\left(R'|A\right)}{L_i Prob(R|A)Prob\left(R'|A\right) + (1 - L_i)Prob(R|A^c)Prob\left(R'|A^c\right)}$$
$$= L_i(R, R')$$

as the responses are independent for persons.

Then, the response-specific jeopardy measure for ith person is defined by

$$J_i(R, R') = \frac{L_i(R, R')/L_i}{(1 - L_i(R, R'))/(1 - L_i)},$$

which indicates the risk of divulging the respondent's status due to the response (R, R'). This $J_i(R, R')$ depends on c_i, unknown probability of direct response.

Usually, the possible responses (R, R') for qualitative ORRTs are (1, 1) (0, 0) (1, 0) and (0, 1). It should be noted that the different responses 1 or 0 coming from the same person for the first and second trials reveal that the person has opted for an RR. But it does not reveal the person's actual status about the stigmatizing characteristic.

A geometric mean (GM) instead of arithmetic mean (AM), earlier suggested by Chaudhuri et al. (2009), is proposed here as an average measure to achieve an algebraic simplicity.

Thus, the proposed measure of jeopardy for ith person is given by

$$\overline{J}_i = \text{G.M of } J_i(R, R') \forall R, R'.$$

In this section, the response-specific measures of jeopardy in the ORR techniques using various RR devices are formulated and combined by the proposed average measure of jeopardy for qualitative characteristics. It can be shown that the measure of jeopardy $\overline{J}_i$ is free from c_i for some well-known ORRTs.

ORR Using Warner's RR Device

Let the response of ith person be (1, 1). Then, using the equation for $L_i(R, R')$, the posterior probability $L_i(1, 1)$ is obtained as follows:

$$L_i(1, 1) = \frac{L_i P(Z_i = 1|y_i = 1)P(Z'_i = 1|y_i = 1)}{L_i P(Z_i = 1|y_i = 1)P(Z'_i = 1|y_i = 1) + (1 - L_i)P(Z_i = 1|y_i = 0)P(Z'_i = 1|y_i = 0)}$$

$$= \frac{L_i\{c_i + (1 - c_i)p_1\}\{c_i + (1 - c_i)p_2\}}{L_i\{c_i + (1 - c_i)p_1\}\{c_i + (1 - c_i)p_2\} + (1 - L_i)\{(1 - c_i)(1 - p_1)\}\{(1 - c_i)(1 - p_2)\}}.$$

With a little algebra,

$$1 - L_i(1, 1)$$
$$= \frac{(1 - L_i)\{(1 - c_i)(1 - p_1)\}\{(1 - c_i)(1 - p_2)\}}{L_i\{c_i + (1 - c_i)p_1\}\{c_i + (1 - c_i)p_2\} + (1 - L_i)\{(1 - c_i)(1 - p_1)\}\{(1 - c_i)(1 - p_2)\}}.$$

Therefore, the response (1, 1)-specific jeopardy measure is given by

$$J_i(1, 1) = \frac{\frac{L_i(1,1)}{L_i}}{\frac{1 - L_i(1,1)}{1 - L_i}} = \frac{\{c_i + (1 - c_i)p_1\}\{c_i + (1 - c_i)p_2\}}{\{(1 - c_i)(1 - p_1)\}\{(1 - c_i)(1 - p_2)\}}$$

using equation for $J_i(R, R')$.

Suppose ith person's response is $(0, 0)$, then the posterior probability $L_i(0, 0)$ is written as

$L_i(0, 0)$

$$= \frac{L_i P(Z_i = 0|y_i = 1)P(Z_i' = 0|y_i = 1)}{L_i P(Z_i = 0|y_i = 1)P(Z_i' = 0|y_i = 1) + (1 - L_i)P(Z_i = 0|y_i = 0)P(Z_i' = 0|y_i = 0)}$$

$$= \frac{L_i\{(1 - c_i)(1 - p_1)\}\{(1 - c_i)(1 - p_2)\}}{L_i\{(1 - c_i)(1 - p_1)\}\{(1 - c_i)(1 - p_2)\} + (1 - L_i)\{c_i + (1 - c_i)p_1\}\{c_i + (1 - c_i)p_2\}}.$$

So,

$$\frac{1 - L_i(0, 0)}{1 - L_i} = \frac{\{c_i + (1 - c_i)p_1\}\{c_i + (1 - c_i)p_2\}}{L_i\{(1 - c_i)(1 - p_1)\}\{(1 - c_i)(1 - p_2)\} + (1 - L_i)\{c_i + (1 - c_i)p_1\}\{c_i + (1 - c_i)p_2\}}.$$

Therefore, the response $(0, 0)$-specific jeopardy measure is written as

$$J_i(0, 0) = \frac{\frac{L_i(0,0)}{L_i}}{\frac{1 - L_i(0,0)}{1 - L_i}} = \frac{\{(1 - c_i)(1 - p_1)\}\{(1 - c_i)(1 - p_2)\}}{\{c_i + (1 - c_i)p_1\}\{c_i + (1 - c_i)p_2\}}.$$

Now for the response $(1, 0)$, the posterior probability $L_i(1, 0)$ and the response-specific jeopardy measure $J_i(1, 0)$ may be written as follows:

$L_i(1, 0)$

$$= \frac{L_i\{c_i + (1 - c_i)p_1\}\{(1 - c_i)(1 - p_2)\}}{L_i\{c_i + (1 - c_i)p_1\}\{(1 - c_i)(1 - p_2)\} + (1 - L_i)\{(1 - c_i)(1 - p_1)\}\{c_i + (1 - c_i)p_2\}},$$

and

$$J_i(1, 0) = \frac{\{c_i + (1 - c_i)p_1\}\{(1 - c_i)(1 - p_2)\}}{\{(1 - c_i)(1 - p_1)\}\{c_i + (1 - c_i)p_2\}}.$$

Similarly, the posterior probability $L_i(0, 1)$ for the response $(0, 1)$ is given by

$L_i(0, 1)$

$$= \frac{L_i\{(1 - c_i)(1 - p_1)\}\{c_i + (1 - c_i)p_2\}}{L_i\{(1 - c_i)(1 - p_1)\}\{c_i + (1 - c_i)p_2\} + (1 - L_i)\{c_i + (1 - c_i)p_1\}\{(1 - c_i)(1 - p_2)\}}.$$

Therefore, the response-specific measure of jeopardy is written as

$$J_i(0, 1) = \frac{\{(1 - c_i)(1 - p_1)\}\{c_i + (1 - c_i)p_2\}}{\{c_i + (1 - c_i)p_1\}\{(1 - c_i)(1 - p_2)\}}.$$

Thus, the proposed measure of jeopardy is the geometric mean of all possible response-specific jeopardy measures, and it is exactly 1 for every individual. There-fore, the overall measure never reveals the status of the respondent. Although $p_1 \rightarrow p_2$ implies that the responses of every individual are well protected but the variance estimate tends to be infinite.

ORR Using Greenberg et al.'s Unrelated Question Model

Now, to calculate posterior probabilities $L_i(R, R')$ and response-specific jeopardy measures $J_i(R, R')$, the necessary conditional probabilities are derived below.

Here,

$$P(Z_i = 1|y_i = 0) = (1 - c_i)(1 - p_1)x_i = (1 - c_i)(1 - p_1),$$

as this happens if the first response of ith sampled person is 1 but the actual value of the stigmatizing characteristic is 0. It is possible only if the sampled person decides to use Greenberg et al.'s (1969) RR device and responds to the question regarding the innocuous characteristic B with the basic assumption that the respondent provide a true response. So, it is obvious that $x_i = 1$.

Thus, $P(Z_i = 0|y_i = 1) = (1 - c_i)(1 - p_1)$.

It is known that $P(A|B) + P(A^C|B) = 1$,

which implies $P(Z_i = 1|y_i = 1) = 1 - P(Z_i = 0|y_i = 1) = c_i + (1 - c_i)p_1$.

and $P(Z_i = 0|y_i = 0) = 1 - P(Z_i = 1|y_i = 0) = c_i + (1 - c_i)p_1$.

Similarly, the conditional probabilities for the second response Z_i' are obtained. Proceeding as above, the posterior probabilities are

$$L_i(1, 1)$$
$$= \frac{L_i\{c_i + (1 - c_i)p_1\}\{c_i + (1 - c_i)p_2\}}{L_i\{c_i + (1 - c_i)p_1\}\{c_i + (1 - c_i)p_2\} + (1 - L_i)\{(1 - c_i)(1 - p_1)\}\{(1 - c_i)(1 - p_2)\}},$$

$$L_i(0, 0)$$
$$= \frac{L_i\{(1 - c_i)(1 - p_1)\}\{(1 - c_i)(1 - p_2)\}}{L_i\{(1 - c_i)(1 - p_1)\}\{(1 - c_i)(1 - p_2)\} + (1 - L_i)\{c_i + (1 - c_i)p_1\}\{c_i + (1 - c_i)p_2\}},$$

$$L_i(1, 0)$$
$$= \frac{L_i\{c_i + (1 - c_i)p_1\}\{(1 - c_i)(1 - p_2)\}}{L_i\{c_i + (1 - c_i)p_1\}\{(1 - c_i)(1 - p_2)\} + (1 - L_i)\{(1 - c_i)(1 - p_1)\}\{c_i + (1 - c_i)p_2\}},$$

and

$$L_i(0, 1)$$
$$= \frac{L_i\{(1 - c_i)(1 - p_1)\}\{c_i + (1 - c_i)p_2\}}{L_i\{(1 - c_i)(1 - p_1)\}\{c_i + (1 - c_i)p_2\} + (1 - L_i)\{c_i + (1 - c_i)p_1\}\{(1 - c_i)(1 - p_2)\}}.$$

Response-specific jeopardy measures are given below.

$$J_i(1, 1) = \frac{\{c_i + (1 - c_i)p_1\}\{c_i + (1 - c_i)p_2\}}{\{(1 - c_i)(1 - p_1)\}\{(1 - c_i)(1 - p_2)\}}$$

$$J_i(0, 0) = \frac{\{(1 - c_i)(1 - p_1)\}\{(1 - c_i)(1 - p_2)\}}{\{c_i + (1 - c_i)p_1\}\{c_i + (1 - c_i)p_2\}}$$

$$J_i(1, 0) = \frac{\{c_i + (1 - c_i)p_1\}\{(1 - c_i)(1 - p_2)\}}{\{(1 - c_i)(1 - p_1)\}\{c_i + (1 - c_i)p_2\}}$$

and

$$J_i(0, 1) = \frac{\{(1 - c_i)(1 - p_1)\}\{c_i + (1 - c_i)p_2\}}{\{c_i + (1 - c_i)p_1\}\{(1 - c_i)(1 - p_2)\}}.$$

Therefore, using the proposed measure of jeopardy.
$\bar{J}_i = \{J_i(1, 1) \times J_i(0, 0) \times J_i(1, 0) \times J_i(0, 1)\}^{1/4} = 1$, whatever be the value of the selection probability of a card from RR devices. Here p_1 cannot tend to p_2, otherwise, the estimate of variance will be infinite.

ORR Using Forced RR Model

Then, the posterior probabilities $L_i(R, R')$ and the response-specific jeopardy measures $J_i(R, R')$ for different responses are as follows:

$L_i(1, 1)$
$$= \frac{L_i\{c_i + (1 - c_i)(1 - p_2)\}\{c_i + (1 - c_i)(1 - p_4)\}}{L_i\{c_i + (1 - c_i)(1 - p_2)\}\{c_i + (1 - c_i)(1 - p_4)\} + (1 - L_i)\{(1 - c_i)p_1\}\{(1 - c_i)p_3\}}$$

$L_i(0, 0)$
$$= \frac{L_i\{(1 - c_i)p_2\}\{(1 - c_i)p_4\}}{L_i\{(1 - c_i)p_2\}\{(1 - c_i)p_4\} + (1 - c_i)\{c_i + (1 - c_i)(1 - p_1)\}\{c_i + (1 - c_i)(1 - p_3)\}}$$

$L_i(1, 0)$
$$= \frac{L_i\{c_i + (1 - c_i)(1 - p_2)\}\{(1 - c_i)p_4\}}{L_i\{c_i + (1 - c_i)(1 - p_2)\}\{(1 - c_i)p_4\} + (1 - L_i)\{(1 - c_i)p_1\}\{c_i + (1 - c_i)(1 - p_3)\}}$$

$L_i(0, 1)$
$$= \frac{L_i\{(1 - c_i)p_2\}\{c_i + (1 - c_i)(1 - p_4)\}}{L_i\{(1 - c_i)p_2\}\{c_i + (1 - c_i)(1 - p_4)\} + (1 - L_i)\{c_i + (1 - c_i)(1 - p_1)\}\{(1 - c_i)p_3\}}$$

$$J_i(1, 1)$$
$$= \frac{\{c_i + (1 - c_i)(1 - p_2)\}\{c_i + (1 - c_i)(1 - p_4)\}}{\{(1 - c_i)p_1\}\{(1 - c_i)p_3\}}$$

$$J_i(0, 0)$$
$$= \frac{\{(1 - c_i)p_2\}\{(1 - c_i)p_4\}}{\{c_i + (1 - c_i)(1 - p_1)\}\{c_i + (1 - c_i)(1 - p_3)\}}$$

$$J_i(1, 0) = \frac{\{c_i + (1 - c_i)(1 - p_2)\}\{(1 - c_i)p_4\}}{\{(1 - c_i)p_1\}\{c_i + (1 - c_i)(1 - p_3)\}}$$

and

$$J_i(0, 1) = \frac{\{(1 - c_i)p_2\}\{c_i + (1 - c_i)(1 - p_4)\}}{\{c_i + (1 - c_i)(1 - p_1)\}\{(1 - c_i)p_3\}}.$$

Therefore, using the proposed jeopardy measure for ith person, it is obtained that

$$\overline{J}_i = \left[\frac{p_2^2 p_4^2 \{c_i + (1 - c_i)(1 - p_2)\}^2 \{c_i + (1 - c_i)(1 - p_4)\}^2}{p_1^2 p_3^2 \{c_i + (1 - c_i)(1 - p_1)\}^2 \{c_i + (1 - c_i)(1 - p_3)\}^2}\right]^{1/4}$$

$$= \frac{p_2}{p_1}\left[\frac{\{c_i + (1 - c_i)(1 - p_2)\}\{c_i + (1 - c_i)(1 - p_4)\}}{\{c_i + (1 - c_i)(1 - p_1)\}\{c_i + (1 - c_i)(1 - p_3)\}}\right]^{1/2}.$$

$$= \frac{p_2}{p_1}\left[\frac{\{c_i + (1 - c_i)(1 - p_2)\}\{c_i + (1 - c_i)(1 - p_4)\}}{\{c_i + (1 - c_i)(1 - p_1)\}\{c_i + (1 - c_i)(1 - p_3)\}}\right]^{1/2}.$$

Here, the geometric mean does not always need to be unity. The overall measure of jeopardy depends on the selection of p_1, p_2, p_3 and p_4. This $\overline{J}_i \rightarrow 1$ if $p_1 \rightarrow p_2$ and $p_3 \rightarrow p_4$.

ORR Using Kuk's RR Model

Considering $(f_i, f_i\prime)$ as the possible responses, posterior probabilities can be written as

$$L_i(f_i, f_i')$$

$$= \frac{L_i P(Z_i = f_i | y_i = 1) P(Z_i' = f_i' | y_i = 1)}{L_i P(Z_i = f_i | y_i = 1) P(Z_i' = f_i' | y_i = 1) + (1 - L_i) L_i P(Z_i = f_i | y_i = 0) P(Z_{i'} = f_i\prime | y_i = 0)}$$

$$= \frac{L_i \psi_{1i} \psi_{1i}'}{L_i \psi_{1i} \psi_{1i}' + (1 - L_i)\psi_{2i} \psi_{2i}'} \quad ; \forall f_i, f_i' = 0, 1, 2 \ldots k,$$

where $\psi_{1i} = c_i I_i + (1 - c_i)\theta_1^{f_i}(1 - \theta_1)^{k - f_i}$ with the indicator function $I_i = \begin{cases} 1, \text{if } f_i = 1 \\ 0 \text{ otherwise} \end{cases}$ and $\psi_{1i}' = c_i I_i' + (1 - c_i)\theta_1^{f_i'}(1 - \theta_1)^{k - f_i'}$ with another indicator function $I_i' = \begin{cases} 1, \text{if } f_i' = 1 \\ 0 \text{ otherwise} \end{cases}$.

Similarly, $\psi_{2i} = c_i I_i + (1 - c_i)\theta_2^{f_i}(1 - \theta_2)^{k - f_i}$ and $\psi_{2i}' = c_i I_i' + (1 - c_i)\theta_2^{f_i'}(1 - \theta_2)^{k - f_i'}$ with two indicator functions defined just above, and the response-specific jeopardy measure $J_i(f_i, f_i')$ is as follows:

$$J_i(f_i, f_i') = \frac{\psi_{1i} \psi_{1i}'}{\psi_{2i} \psi_{2i}'} = J_i(f_i).J_i(f_i'),$$

where $J_i(f_i) = \frac{\psi_{1i}}{\psi_{2i}}; J_i(f_i') = \frac{\psi_{1i}'}{\psi_{2i}'} \; \forall f_i, f_i' = 0, 1, 2 \ldots k,$
and

$$\overline{J}_i = \prod_{\forall f_i, f_i'} (J_i(f_i, f_i'))^{1/(k+1)^2} = \prod_{\forall f_i, f_i'} (J_i(f_i)J_i(f_i'))^{1/(k+1)^2} = \prod_{\forall f_i} (J_i(f_i))^{1/(k+1)}.$$

Consequently, $J_i(0) = \frac{(1-c_i)(1-\theta_1)^k}{c_i+(1-c_i)(1-\theta_2)^k}$ and $J_i(1) = \frac{c_i+(1-c_i)\theta_1(1-\theta_1)^{k-1}}{(1-c_i)\theta_2(1-\theta_2)^{k-1}}$ do not tend to 1 whatever the choice of θ_1, θ_2.

But, $J_i(f_i) = \frac{(1-c_i)\theta_1^{f_i}(1-\theta_1)^{k-f_i}}{(1-c_i)\theta_2^{f_i}(1-\theta_2)^{k-f_i}} = \left(\frac{\theta_1}{\theta_2}\right)^{f_i}\left(\frac{1-\theta_1}{1-\theta_2}\right)^{k-f_i}$, for all $f_i, f_i' = 2, 3 \ldots k,$
and it tends to 1 if $\theta_1 \to \theta_2$.

$$\overline{J}_i = [J_i(0) \cdot J_i(1) \cdot J_i(2) \ldots J_i(k-1) \cdot J_i(k)]^{1/k+1}$$

$$= \left[\frac{(1-c_i)(1-\theta_1)^k}{c_i+(1-c_i)(1-\theta_2)^k} \frac{c_i+(1-c_i)\theta_1(1-\theta_1)^{k-1}}{(1-c_i)\theta_2(1-\theta_2)^{k-1}} \cdot \frac{\theta_1^2(1-\theta_1)^{k-2}}{\theta_2^2(1-\theta_2)^{k-2}} \frac{\theta_1^3(1-\theta_1)^{k-3}}{\theta_2^3(1-\theta_2)^{k-3}} \cdots \frac{\theta_1^k}{\theta_2^k}\right]^{1/k+1}$$

$$= \left[\frac{c_i+(1-c_i)\theta_1(1-\theta_1)^{k-1}}{c_i+(1-c_i)(1-\theta_2)^k} \frac{(1-c_i)(1-\theta_1)^k}{(1-c_i)\theta_2(1-\theta_2)^{k-1}} \cdot \frac{\theta_1^2(1-\theta_1)^{k-2}}{\theta_2^2(1-\theta_2)^{k-2}} \frac{\theta_1^3(1-\theta_1)^{k-3}}{\theta_2^3(1-\theta_2)^{k-3}} \cdots \frac{\theta_1^k}{\theta_2^k}\right]^{1/k+1}.$$

It is observed that in the case of ORR with Kuk's RRT, the overall measure of jeopardy $\overline{J}_i$ tends to 1 if $\theta_1, \theta_2 \to \frac{1}{2}$.

Numerical studies have been undertaken from Pal et al. (2020)—the work by all the three authors of this monograph, to demonstrate the proposed method of protecting the privacy of the responses employing various ORR models. Prior probabilities (L_i) and the probabilities of direct response (c_i) were considered there to be known, which were actually unknown in practice. We have included the same numerical in Tables 7.1, 7.2 and 7.3 taking different combinations of (L_i, c_i) values to calculate the posterior probabilities, and therefore, the response-specific jeopardy measures $J_i(R, R')$. The overall measure of jeopardy $(\overline{J}_i)$ is shown in the last column of the mentioned tables.

Table 7.1 ORRT with Warner's and Greenberg et al.'s model—measure of jeopardy

L_i	c_i	p_1	p_2	$J_i(0,1)$	$J_i(1,0)$	$J_i(0,0)$	$J_i(1,1)$	Warner's $\overline{J}_i$	Unrelated $\overline{J}_i$
0.1	0.6	0.44	0.49	1.2216	0.8186	1.0409	0.9607	1	1
0.3	0.63	0.3	0.73	3.1622	0.3162	0.039	25.6154	0.9989	0.9989
0.4	0.42	0.95	0.11	0.0285	35.028	0.0335	29.8462	0.9981	0.9981
0.5	0.91	0.42	0.28	0.8246	1.2128	0.0034	297.6667	1.0121	1.0121
0.6	0.37	0.07	0.98	142.46	0.007	0.0145	68.7966	0.9948	0.9948
0.8	0.55	0.43	0.62	1.7154	0.5829	0.072	13.8959	1.0004	1.0004
0.9	0.53	0.57	0.73	1.6731	0.5977	0.0374	26.7692	1.0012	1.0012

Table 7.2 ORRT with Boruch's model—Measure of jeopardy

L_i	c_i	p_1	p_2	p_3	p_4	$J_i(1, 0)$	$J_i(0, 1)$	$J_i(0, 0)$	$J_i(1, 1)$	Forced $\overline{J_i}$
0.1	0.42	0.64	0.23	0.24	0.09	0.1367	1.4002	0.012	15.9556	0.4375
0.2	0.43	0.45	0.4	0.52	0.462	1.1	0.7667	0.1154	7.3051	0.9183
0.6	0.18	0.61	0.25	0.47	0.192	0.4195	0.8615	0.1049	3.4467	0.6013
0.6	0.26	0.39	0.31	0.43	0.342	0.9762	0.7592	0.1191	6.2231	0.8609
0.7	0.3	0.25	0.4	0.37	0.592	2.2162	0.7749	0.1892	9.0769	1.3105
0.9	0.1	0.69	0.21	0.6	0.183	0.4544	0.7778	0.1739	2.0323	0.5945
0.9	0.12	0.58	0.35	0.37	0.223	0.4039	1.5337	0.1889	3.2799	0.7871

Table 7.3 ORRT with Kuk's model—measure of jeopardy

L_i	c_i	θ_1	θ_2	$J_i(0)$	$J_i(1)$	$J_i(2)$	Kuk $\overline{J_i}$
0.1	0.18	0.72	0.43	0.1333	1.75	2.867	0.874
0.2	0.36	0.78	0.13	0.0357	6.7143	39	2.107
0.3	0.55	0.54	0.34	0.1333	6.6	2.6	1.318
0.4	0.72	0.49	0.5	0.0886	11.286	1	1
0.5	0.8	0.58	0.53	0.0476	17	1.167	0.981
0.6	0.74	0.55	0.2	0.0549	20	8	2.063
0.7	0.78	0.75	0.76	0.0127	20.5	0.923	0.622
0.9	0.49	0.77	0.81	0.0588	7.25	0.909	0.729

　　As noted in Sect. 7.4.1 for ORRTs with Warner's RRT and Greenberg et al.'s RRT, $\overline{J_i}$ is exactly 1 whatever the value of L_i, c_i, p_1 and p_2, but here in the numerical illustration for the ORR techniques (see Table 7.1), it is slightly different from 1 because of the approximation to the four decimal places in the calculation of posterior probabilities and response-specific jeopardy measures. Table 7.2 computes the measures for ORR using Boruch's (1971) RR device restricting $p_1 p_4 = p_2 p_3$ as pointed out in Sect. 7.4.1. Table 7.3 shows the numerical study of protection of privacy using the ORR technique with Kuk's (1990) RR device. Here, it is considered that if a respondent opts for Kuk's RR device, he/she should draw two cards (i.e. $k = 2$) from the RR device.

7.5 Protection of Privacy Measures for Quantitative ORRT

7.5.1 Patra and Pal (2019)'s Privacy Measure (Quantitative)

Considering $P(Z_i|y_i)$ as the conditional probability of the reported ORR value for ith person while the true response is y_i, Chaudhuri and Christofides (2013)'s measure of the jeopardy is extended here for quantitative ORRTs. It has been shown that $L_i(y_i|Z_i) = L_i$, i.e. the posterior probability is equal to the prior probability, for ORRTs with Eichhorn and Hayre's RR device and also for Chaudhuri's RR device I.

For ORR using Eichhorn and Hayre's RR device,

$$P(Z_i|y_i) = c_i + (1-c_i)\left(\sum_{x_j} P(x = x_j)P\left(\psi = Z_i - \frac{y_i x_j}{\theta_1}\right)\right) = P(Z_i),$$

and for ORR using Chaudhuri's RR device I,

$$P(Z_i|y_i) = c_i + \frac{1}{TM}(1-c_i) = P(Z_i)$$

Thus, privacy is well protected for each individual by the ORR methods with the mentioned RR devices.

However, such conclusion about the posterior and prior probabilities is not true for the ORRT using Chaudhuri's RR device II as

$$L_i(y_i|z_i) = \frac{L_i(c_i + k(1-c_i))}{L_i(c_i + k(1-c_i)) + (1-L_i)(1-c_i)(1-k)} = \left[1 + \frac{1-L_i}{L_i}\frac{\theta_i}{1-\theta_i}\right]^{-1},$$

which depends on unknown c_i. Here, $\theta_i = (1-c_i)(1-k)$ and $L_i(y_i|Z_i)$ approaches L_i if and only if $\theta_i \to \frac{1}{2}$ and this θ_i is unknown.

Some numerical illustrations are presented here from the unpublished thesis by one of the authors of this monograph (Patra (2022)) to study the effectiveness and competitiveness of the proposed quantitative ORR models. Tables 7.4, 7.5 and 7.6 provide the measure of jeopardy along with the posterior probabilities for the different combinations of prior probabilities and DR probabilities (L_i, c_i). Table 7.4 shows the numerical for the privacy measure of the quantitative ORR model with the RR device of Eichhorn and Hayre (1983). Similarly, Tables 7.5 and 7.6 represent the numericals for the privacy measure of ORR devices with Chaudhuri (2011a, 2011b)'s RR devices I and II as described in Sect. 7.5.1.

Table 7.4 Privacy measure for ORR with Eichhorn and Hayre (RR) device

L_i	c_i	θ_1	$L(y_i|z_i)$	Measure of Jeopardy
0.15	0.22	13	0.15	1
0.26	0.36	13	0.26	1
0.39	0.68	12	0.39	1
0.54	0.43	12	0.54	1
0.64	0.21	13	0.64	1
0.82	0.31	11	0.82	1
0.88	0.64	14	0.88	1

Table 7.5 Privacy measure for ORR with Chaudhuri's RR device I

L_i	c_i	μ_b	$L(y_i|z_i)$	Measure of Jeopardy
0.15	0.22	2	0.15	1
0.26	0.36	7	0.26	1
0.39	0.68	19	0.39	1
0.54	0.43	8	0.54	1
0.64	0.21	13	0.64	1
0.82	0.31	12	0.82	1
0.88	0.64	3	0.88	1

Table 7.6 Privacy measure for ORR with Chaudhuri's RR device II

L_i	c_i	k	$L(y_i|z_i)$	Measure of Jeopardy
0.20	0.29	0.36	0.23055	1.15275
0.26	0.65	0.89	0.35135	1.35135
0.56	0.33	0.49	0.71031	1.26841
0.64	0.8	0.34	0.92119	1.43937
0.74	0.2	0.37	0.84955	1.14805
0.82	0.37	0.3	0.85239	1.03949
0.9	0.29	0.16	0.85897	0.9544

References

Anderson, H. (1975a). *Efficiency versus protection in the RR for estimating proportions.* Technical Report 9, University of Lund, Lund, Sweden.

Anderson, H. (1975b). *Efficiency versus protection in a general RR model.* Technical Report 10, University of Lund, Lund, Sweden.

Anderson, H. (1975c). *Efficiency versus protection in RR designs.* University of Lund, Lund, Sweden.

Anderson, H. (1977). Efficiency versus protection in a general randomized response model. *Scandinavian Journal of Statistics, 4*, 11–19.

Boruch, R. F. (1971). Assuring confidentiality of responses in social research: A note on strategies. *American Sociologist, 6*, 308–311.

Bose, M. (2015). Respondent privacy and estimation efficiency in randomized response surveys for discrete-valued sensitive variables. *Statistical Papers, 56*, 1055–1069.

Bose, M., & Dihidar, K. (2018). Privacy protection measures for randomized response surveys on stigmatizing continuous variables. *Journal of Applied Statistics*.

Chaudhuri, A. (2001). Estimating sensitive proportions from unequal probability samples using randomized responses. *Pakistan Journal of Statistics, 17*, 259–270.

Chaudhuri, A. (2011a). *Randomized response and indirect questioning techniques in surveys*. CRC Press.

Chaudhuri, A. (2011a). Unequal probability sampling: Analyzing optional randomized response on qualitative and quantitative variables bearing social stigma. In: *International Statistical Institute Proceedings 58th World Statistical Congress, 2011,*, (pp. 1939–1947). Dublin.

Chaudhuri, A., & Christofides, T. C. (2013). *Indirect questioning in sample surveys*. Springer Verlag.

Chaudhuri, A., & Mukerjee, R. (1985). Optionally randomized responses techniques. *Calcutta Statistical Association Bulletin, 34*, 225–229.

Chaudhuri, A., & Saha, A. (2004). Utilizing covariates by logistic regression modelling in improved estimation of population proportions bearing stigmatizing features through random-ized responses in complex surveys. *Journal of Indian Society of Agricultural Statistics, 58(2)*, 190–211.

Chaudhuri, A., Christofides, T. C., & Saha, A. (2009). Protection of privacy in efficient application of randomized response techniques. *Statistical Methods and Applications, 18*, 389–418.

Diana, G., & Perri, P. F. (2008). Efficiency vs privacy protection in SRR methods. In *Proceedings of 44th Scientific Meeting of the Italian Statistical Society*.

Diana, G., & Perri, P. F. (2011). A class of estimators for quantitative sensitive data. *Statistical Papers, 52(3)*, 633–650.

Diana, G., Giordan, M., & Perii, P. F. (2013). Randomized response surveys: A note on some privacy protection measures. *Model Assisted Statistics and Applications, 8(1)*, 19–28.

Eichhorn, B., & Hayre, L. S. (1983). Scrambled randomized response methods for obtaining sensitive quantitative data. *Journal of Statistical Planning and Inference, 7*, 307–316.

Fligner, M. A., Policello, G. E., & Singh, J. (1977). A comparison of two randomized response survey methods with consideration for the level of respondent protection. *Communications in Statistics-Theory and Methods, 6(15)*, 1511–1524.

Greenberg, B. G., Abul-Ela, A. L., Simmons, W. R., & Horvitz, D. G. (1969). The unrelated question randomized response model: Theoretical framework. *Journal of American Statistical Association, 64*, 520–539.

Gupta, S., Mehta, S., Shabbir, J., & Khalil, S. (2018). A unified measure of respondent privacy and model efficiency in quantitative RRT models. *Journal of Statistical Theory and Practice, 12*, 506–511.

Kuk, A. Y. (1990). Asking sensitive questions indirectly. *Biometrika, 77(2)*, 436–438.

Lanke, J. (1975). On the choice of the unrelated question in Simmons' version of randomized response. *Journal of American Statistical Association, 70*, 80–83.

Lanke, J. (1976). On the degree of protection in randomized interviews. *International Statistical Review, 44*, 197–203.

Leysieffer, R. W., & Warner, S. L. (1976). Respondent jeopardy and optimal designs in randomized response models. *Journal of American Statistical Association, 71*, 649–656.

Nayak, T. K. (1994). On randomized response surveys for estimating a proportions. *Communications in Statistics—Theory and Methods, 23*, 3303–3321.

Pal, S., Chaudhuri, A., & Patra, D. (2020). How privacy may be protected in optional randomized response surveys. *Statistics in Transition, 21(2)*, 61–87.

Patra, D. (2022). *Investgating methods of finite population sampling with varying probabilities*. (Unpublished Thesis). West Bengal State University.

Patra, D., & Pal, S. (2019). Privacy protection in optional randomized response surveys for quantitative characteristics. *Staistics and Applications, 17*(2), 47–54.

Pollock, K. H., & Bek, Y. (1976). A comparison of three randomized response models for quantitative data. *Journal of American Statistical Association, 71*, 884–886.

Warner, S. L. (1965). Randomized response: A survey technique for eliminating evasive answer bias. *Journal of American Statistical Association, 60*, 63–69.

Yan, Z., Wang, J., & Lai, J. (2008). An efficiency and protection degree based comparison among the quantitative randomized response strategies. *Communications in Statistics- Theory and Methods, 38*(3), 400–408.

Zhimin, H., Zaizai, Y., & Lidong, W. (2010). A note of proposed privacy measures in randomized response models. *Combining soft computing and statistical methods in data analysis. Advances in intelligent and soft computing* (pp. 635–642). Springer, Berlin, Heidelberg.

Chapter 8
Variation from Classical Data Generating Procedures by Repeated Drawing

8.1 Introduction

Consider y_i takes value 1 if the $i^{th} (i = 1, 2, \ldots, N)$ individual in the population U bears a sensitive characteristic A and value 0 if individual i bears A^C. The proportion of individuals $\theta = \frac{1}{N} \sum_{i=1}^{N} y_i$ in the population bearing A is estimated by employing randomized response (RR) devices pioneered by Warner (1965) followed by several other devices existing in literature. These devices usually mandate a sampled respondent to select card(s) randomly from a box answer a 'match' or 'mismatch' according to the characteristic (A or A^C) marked in the selected card(s). Singh and Grewal (2013) demonstrated that alternatively if cards are continued to be drawn from the box until a 'match' is obtained and the total number of trials to get a 'match' is taken as the randomized response of the respondent, then this approach of utilizing inverse Bernoulli trials in generating RRs enhances the accuracy in the unbiased estimation of θ. This approach was later extended to samples drawn by an unequal probability sampling design, by Chaudhuri and Dihidar (2014). But Chaudhuri and Shaw (2016) observed that the alternative approach of executing the inverse Bernoulli trials does not work with the traditional Simmons' Unrelated Characteristic Model (URL) which is used when both A and A^C are sensitive. The URL indirectly enquires about A; as A^C is also sensitive, it is replaced by an innocuous characteristic, say B with values x_i which is 1, if i bears B and 0, else. B may denote love for reading books more than for sport; both B and B^C should be unrelated to A and A^C. Chaudhuri's (2011) version requires the selection of one sample, instead of two independent samples and is applicable to samples chosen by unequal probability sampling design. A sampled person i $(i \in s)$ is offered two boxes, one containing a number of cards marked A and B in proportions p_1, $(0 < p_1 < 1)$ and $(1 - p_1)$, respectively, and the other containing numerous cards marked A and B in proportions p_2, $(0 < p_2 < 1)$ and $(1 - p_2)$, $p_1 \neq p_2$. Then, the person independently and randomly chooses one card from the first box and another card from the second box and answers a 'match' or 'mismatch' for the first and second cards. Chaudhuri

© The Author(s), under exclusive license to Springer Nature Singapore Pte Ltd. 2024
A. Chaudhuri et al., *Randomized Response Techniques*,
https://doi.org/10.1007/978-981-99-9669-8_8

and Shaw (2016) found that application of inverse Bernoulli trials in drawing cards from the boxes may result in infinite number of draws for an individual as he/she may neither bear A nor B. Hence, they modified the URL first and then incorporate inverse Bernoulli trials to the modified version.

On the other hand, Singh and Sedory (2013) studied the generation of RRs by using inverse Hypergeometric trials, i.e. reporting the number of trials performed to achieve a particular number of successes. Dihidar (2016) presented a follow-up paper by generalizing the sampling design of drawing the sample from the population. Her findings showed that RRs generated by following the inverse Hypergeometric trials are more efficient than those obtained by following the direct Hypergeometric trials. Motivated by these, Shaw and Chaudhuri (2022) have made an attempt to improve the revised version of Greenberg et al.'s (1971) URL RRT by modifying the device so as to implement both direct and inverse Hypergeometric trials.

8.2 Repeated Draws with or Without Replacement

8.2.1 Singh and Grewal's (2013) Approach—Drawing Sample with Replacement

Singh and Grewal (2013) presented a modification in the fashion of draws from the urns in the Kuk's (1990) RRT. In Kuk's (1990) RRT, a sampled individual $i (i \in s)$ is offered two boxes, one containing red identical cards in proportion p_1 and non-red identical cards in proportion $(1 - p_1)$ and another containing red identical cards in proportion $p_2(\neq p_1)$ and non-red identical cards in proportion $(1 - p_2)$. The individual is requested to draw K cards with replacement from the first box, if he/she bears A, else from the second box. Without disclosing the chosen box, the individual i is requested to answer the number of red cards (say f_i) out of the K cards drawn. It is already discussed in Sect. 2.5 of Chap. 2.

Singh and Grewal (2013) presented a modification in the fashion of draws from the urns in the Kuk's (1990) RRT. They demonstrated that alternatively if cards are continued to be drawn from a box until a 'match' is obtained and the total number of trials to get a 'match' is taken as the RR of the respondent, then this approach of utilizing inverse Bernoulli trials in generating RR's enhances the accuracy in the unbiased estimation of θ. However, they restricted this method for samples drawn from the population by using SRSWR.

In their model, if a respondent belongs to group A, he/she is instructed to draw cards, one-by-one using *with replacement,* from the first deck of cards until he/she gets the first card bearing the statement of his/her own status, and requested to report the total number of cards, say X, drawn by him/her to obtain the first card of his/her own status.

The details of the estimation process are given in Sect. 10.3 (in Chap. 10) of this book.

8.2.2 Chaudhuri and Dihidar's (2014) Approach—General Sampling Design

Chaudhuri and Dihidar (2014) extended Singh and Grewal (2013)'s approach to Warner's (1965) and Kuk's (1990) RRT while allowing samples to be chosen from the population by a general sampling scheme. Following the pioneering work of Singh and Grewal (2013), they examined how their inverse approach of revising Warner's and Kuk's techniques of directly eliciting a randomized response to unbiasedly estimate a finite population proportion bearing a stigmatizing feature, fares as a viable competitor.

Applying Singh and Grewal's (2013) approach in Warner's RRT, let a sampled person labelled i ($i \in s$) be requested to continue drawing randomly with replacement one card from the box and report the value g_i which is the number of draws on which first time a 'Match' in card type and the person's feature be observed. The sample s is suitably chosen with a probability $p(s)$ according to a design. Then recalling from Walpole and Myers (1993) the mean and variance of the number of draws needed for the first success in Bernoulli trials with p as the probability of 'success' it follows that

$$E_R(g_i) = \frac{y_i}{p} + \frac{1 - y_i}{1 - p} \text{ and}$$

$$V_R(g_i) = y_i \frac{1 - p}{p^2} + (1 - y_i) \frac{p}{(1 - p)^2}, i = 1, 2, \ldots, N.$$

Here, y_i is unbiasedly estimated by $r_i = \frac{g_i - (1-p)^{-1}}{p^{-1} - (1-p)^{-1}} = \frac{p(1-p)g_i - p}{(1-2p)}$

with variance $V(r_i) = V_i = [\frac{p(1-p)}{(1-2p)}]^2 V_R(g_i) = \frac{p^3}{(1-2p)^2} + \frac{1 - p + p^2}{1 - 2p} y_i.$

So $E_R(r_i) = y_i$.

An unbiased estimator for V_i is $\widehat{V_i} = \frac{p^3}{(1-2p)^2} + \frac{1-p+p^2}{1-2p} r_i.$

As per Chaudhuri and Dihidar (2014), in case of Kuk's (1990) RRT, a sampled person labelled i ($i \in s$) using the appropriate box as instructed is to report the number K denoting the draw on which he/she gets on repeated trials with replacement the 'red' card for the first time.

The number K_i denotes the draw on which he/she gets on repeated trials with replacement the 'red' card for the first time.

$$E_R(K_i) = \frac{y_i}{p_1} + \frac{1 - y_i}{p_2}, \text{ and}$$

$$V_R(K_i) = \frac{y_i(1 - p_1)}{p_1^2} + \frac{(1 - y_i)(1 - p_2)}{p_2^2}.$$

The terms y_i and $V_R(K_i)$ are unbiasedly estimated by $\widehat{y_i} = r_i = \frac{k_i - \frac{1}{p_2}}{p_2 - p_1}$, $(p_2 \neq p_1)$,

and $\widehat{V_R(K_i)} = v_i$, where

$$v_i = \frac{p_1^2 p_2^2}{(p_2 - p_1)^2} [\frac{1 - p_2}{(p_2)^2} + \frac{(p_2 - p_1)(p_1 + p_2 - p_1 p_2)}{p_1^2 p_2^2}] r_i.$$

Clearly, $E_R(\widehat{y_i}) = E_R(r_i) = y_i$ and $E_R(v_i) = V_R(K_i), i \in s$.

8.2.3   Singh and Sedory's (2013) Approach

On the other hand, Singh and Sedory (2013) studied the generation of RRs by using inverse Hypergeometric trials, i.e. reporting the number of trials performed to achieve a particular number of successes.

According to their model, randomized response device consists of two urns: Urn-I and Urn-II.

Urn-I contains N_1 balls, out of which r_1 balls bearing the statement, (i) "I belong to the group A", and the remaining $(N_1 - r_1)$ balls are blank with no statement on them.

Urn-II contains N_2 balls, out of which r_2 balls bearing the statement, (ii) "I do not belong to the group A", and the remaining $(N_2 - r_2)$ balls are blank with no statement on them. Each respondent selected in the sample is instructed as follows: If he/she belongs to the group A, then he/she draws balls using without replacement sampling from the Urn-I until he/ she gets t_1 $(< r_1)$ balls bearing the statement (i) and reports the total number of balls, say X, drawn by him/her. Thus, X follows a negative hypergeometric distribution which is given by

$$P(X = x | N_1, r_1, t_1) = \frac{\binom{x - 1}{t_1 - 1}\binom{N_1 - x}{r_1 - t_1}}{\binom{N_1}{r_1}}, x = t_1, t_1 + 1, \ldots \ldots, N_1 - r_1 + t_1.$$

If he/she does not belongs to the group A, then he/she draws balls using without replacement sampling from the Urn-II until he/she gets t_2 $(< r_2)$ balls bearing the statement (ii) and reports the total number of balls, say Y, drawn by him/her. Thus, Y also follows a negative hypergeometric distribution such as

$$P(Y = y | N_2, r_2, t_2) = \frac{\binom{y-1}{t_2-1}\binom{N_2-y}{r_2-t_2}}{\binom{N_2}{r_2}}, \; y = t_2, t_2 + 1, \ldots\ldots, N_2 - r_2 + t_2.$$

The response of ith ($i \in s$) respondent is

$$\begin{aligned} Z_i &= X \; if \; i \in A \\ &= Y \; if \; i \in A^c. \end{aligned} \quad (8.1)$$

The population proportion θ is unbiasedly estimated by

$$\hat{\theta} = \frac{\frac{(r_1+1)(r_2+1)}{n} \sum_1^n Z_i - t_2(r_1+1)(N_2+1)}{t_1(N_1+1)(r_2+1) - t_2(N_2+1)(r_1+1)}.$$

The variance is

$$V(\hat{\theta}) = \frac{\frac{[(r_1+1)(r_2+1)]^2}{n}(\sigma_Z)^2}{[t_1(N_1+1)(r_2+1) - t_2(N_2+1)(r_1+1)]^2}.$$

8.2.4 Dihidar's (2016) Modification on Singh and Sedory's Work for General Sampling Design

Dihidar (2016) presented a follow-up paper on Singh and Sedory (2013) by generalizing the sampling design of drawing the sample from the population. Her findings showed that RRs generated by following the inverse Hypergeometric trials are more efficient than those obtained by following the direct Hypergeometric trials. First they presented the deviation for generating RR by direct Hypergeometric distribution.

In direct Hypergeometric distribution scheme, two boxes are given to the respondents. The first box contains total N_1 cards, out of which r_1 cards are red and $N_1 - r_1$ cards are blue. The second box contains N_2 cards, out of which r_2 cards are red and the rest ($N_2 - r_2$) cards are blue. Here, $\frac{r_1}{N_1} \neq \frac{r_2}{N_2}$.

A sample s of size n is drawn from the population with a general sampling design $p(s)$. Each sampled respondent is requested to draw cards K times without replacement either from first box or second box according to whether he/she bears the sensitive characteristic A or A^C. . Each respondent will report the number of red cards (say f_i) drawn out of K cards.

We can write

$$E_R(f_i) = K\left[y_i \frac{r_1}{N_1} + (1 - y_i)\frac{r_2}{N_2}\right] \text{ vide Chaudhuri (2001)'s approach,}$$

and

$$V_R(f_i) = K\left[y_i \frac{r_1}{N_1} \frac{N_1 - r_1}{N_1} \frac{N_1 - K}{N_1 - 1} + (1 - y_i)\frac{r_2}{N_2} \frac{N_2 - r_2}{N_2} \frac{N_2 - K}{N_2 - 1}\right].$$

Writing $r_i = \frac{\frac{f_i}{K} - \frac{r_2}{N_2}}{\frac{r_1}{N_1} - \frac{r_2}{N_2}}$, we have $E_R(r_i) = y_i$.

$$V_R(r_i) = \frac{\frac{V_R(f_i)}{K^2}}{(\frac{r_1}{N_1} - \frac{r_2}{N_2})^2} = ay_i + b.$$

Here,

$$a = \frac{1}{K(\frac{r_1}{N_1} - \frac{r_2}{N_2})^2}\left[\frac{r_1}{N_1} \frac{N_1 - r_1}{N_1} \frac{N_1 - K}{N_1 - 1} - \frac{r_2}{N_2} \frac{N_2 - r_2}{N_2} \frac{N_2 - K}{N_2 - 1}\right], \text{ and}$$

$$b = \frac{1}{K(\frac{r_1}{N_1} - \frac{r_2}{N_2})^2}\left[\frac{r_2}{N_2} \frac{N_2 - r_2}{N_2} \frac{N_2 - K}{N_2 - 1}\right].$$

The term $V_R(r_i)$ is estimated by $v_i = a + br_i$. Here, $E_R(v_i) = V_R(r_i)$.

Dihidar (2016) also considered negative hypergeometric distribution to generate RR from respondents, drawing sample by general sampling design. The RR response is given in (8.1).

The expectation and variance of the negative hypergeometric distribution are

$$E(X) = t_1 \frac{N_1 + 1}{r_1 + 1}, \text{ and}$$

$$V(X) = t_1 \frac{(N_1 - r_1)(N_1 + 1)(r_1 + 1 - t_1)}{(r_1 + 1)^2(r_1 + 2)}.$$

Similarly, the expectation and variance of the variable Y can be written. So, we may write

$$E_R(Z_i) = y_i E_R(X) + (1 - y_i)E_R(Y).$$

So, y_i is unbiasedly estimated by

$$r_i^* = \frac{Z_i - t_2 \frac{N_2 + 1}{r_2 + 1}}{\frac{t_1(N_1 + 1)(r_2 + 1) - t_2(N_2 + 1)(r_1 + 1)}{(r_1 + 1)(r_2 + 1)}}.$$

It implies $E_R(r_i^*) = y_i$.

$$V_R(r_i^*) = \frac{V_R(Z_i)}{[\frac{t_1(N_1 + 1)(r_2 + 1) - t_2(N_2 + 1)(r_1 + 1)}{(r_1 + 1)(r_2 + 1)}]^2},$$

$$V_R(Z_i) = V_i^* = E_R\left(Z_i^2\right) - (E_R(Z_i))^2$$
$$= y_i E_R\left(X^2\right) + (1 - y_i)E_R\left(Y^2\right) - (E_R(Z_i))^2$$
$$= y_i\{V_R(X) + E_R(X)\}^2 + (1 - y_i)\{V_R(Y) + E_R(Y)\}^2 - (E_R(Z_i))^2$$
$$= cy_i + d, \text{ where}$$

$$c = \dfrac{\dfrac{t_1(N_1-r_1)(N_1+1)(r_1+1-t_1)}{(r_1+1)^2(r_1+2)} - \dfrac{t_2(N_2-r_2)(N_2+1)(r_2+1-t_2)}{(r_2+1)^2(r_2+2)}}{[\dfrac{t_1(N_1+1)}{r_1+1} - \dfrac{t_2(N_2+1)}{r_2+1}]^2}.$$

V_i^* is unbiasedly estimated by $\widehat{V_i^*} = cr_i + d$.

Motivated by these, Chaudhuri and Shaw (2016) have made an attempt to improve the revised version of Greenberg et al.'s (1971) URL RRT by modifying the device so as to implement both direct and inverse Hypergeometric trials. The details are narrated in Sect. 10.3 of Chap. 10 of this book.

In addition to E_R and V_R, let E_P and V_P denote the design-based expectation and variance operators, respectively, and $E = E_P E_R = E_R E_P$ and $V = E_P V_R + V_P E_R = E_R V_P + V_R E_P$, the overall expectation and variance operators.

Consider the estimator e, where

$$e = \frac{1}{N} \sum_{i \in s} \frac{r_i}{\pi_i}.$$

Then,

$$E(e) = E_R E_P(e) = E_P E_R(e) = \frac{1}{N} \sum_{i=1}^{N} y_i = \theta.$$

Hence, e is an unbiased estimator of θ. Now, using Chaudhuri and Pal (2002), variance of e can be expressed as

$$V(e) = V\left(\frac{1}{N} \sum_{i \in s} \frac{r_i}{\pi_i}\right) = E_R V_P\left(\frac{1}{N} \sum_{i \in s} \frac{r_i}{\pi_i}\right) + V_R E_P\left(\frac{1}{N} \sum_{i \in s} \frac{r_i}{\pi_i}\right)$$

$$= E_R\left[\frac{1}{N^2}\left\{\sum_{i<j=1}^{N}\sum^{N}(\pi_i\pi_j - \pi_{ij})\left(\frac{r_i}{\pi_i} - \frac{r_j}{\pi_j}\right)^2 + \sum_{i=1}^{N} \frac{\beta_i}{\pi_i}r_i^2\right\}\right] + \frac{1}{N^2} \sum_{i=1}^{N} V_R(r_i)$$

$$\beta_i = 1 + \frac{1}{\pi_i} \sum_{j \neq i}^{N} \pi_{ij} - \sum_{i=1}^{N} \pi_i.$$

If every sample s contains a common number of distinct units in it, then $\beta_i = 0 \forall\ i$ throughout in $V(e)$. An unbiased estimator for $V(e)$ is

$$v(e) = \frac{1}{N^2} \left\{ \sum_{i<} \sum_{j \in s} \left(\frac{\pi_i \pi_j - \pi_{ij}}{\pi_{ij}} \right) \left(\frac{r_i}{\pi_i} - \frac{r_j}{\pi_j} \right) + \sum_{j \in s} \frac{\beta_i}{\pi_i^2} r_i^2 \right\} + \frac{1}{N^2} \sum_{j \in s} \frac{V_R(r_i)}{\pi_i}.$$

with $\beta_i = 0 \forall i$ in $v(e)$ when applicable, such that
$E\{v(e)\} = E_P E_R\{v(e)\} = E_R E_P\{v(e)\} = V(e)$. A $100(1 - \alpha)\%$ confidence interval (CI) for θ is, $[L, M]$, where

$$L = e - \left(\tau_{\frac{\alpha}{2}} \sqrt{v(e)} \right), \quad M = e + \left(\tau_{\frac{\alpha}{2}} \sqrt{v(e)} \right),$$

and $\tau_{\alpha/2}$ is the upper $\alpha/2$ point of $N(0, 1)$ distribution.

8.2.5 Generating Randomized Response by Inverse Bernoulli Trials

In Warner's RR model, usually a box is used containing cards bearing either of two types and drawing such cards randomly and reporting match or mismatch of the respondent's true feature with the type of the card actually drawn. Singh and Grewal (2013) introduced an amendment to it by demanding repeated draws of cards from the box and reporting the particular number of draws on which a match is noticed for the first time. This leads to the study of inverse Bernoulli distribution; this achieves higher efficiency in estimation as shown by them followed by Chaudhuri and Dihidar (2014) considering specific RRT.

But Chaudhuri and Shaw (2016) observed that the alternative approach of executing the inverse Bernoulli trials does not work with the traditional URL RRT which is used when both A and A^C are sensitive. They modified the URL RRT first and then incorporate inverse Bernoulli trials to the modified version. To apply Singh and Grewal's (2013) approach of using inverse Bernoulli trials in URL, Chaudhuri and Shaw (2016) found that drawing cards randomly with replacement from an urn till there is a match may not result in a finite number of trials for an individual, as he/ she may neither bear A nor B. Hence, they first presented a modified version of the URL with direct trials for drawing cards and this was further revised incorporating the inverse Bernoulli trials.

Defining w_i as 1, if i bears $(A \cup B)$ and 0, otherwise and u_i as 1, if i bears $(B \cap A^C)$ and 0, otherwise, thus giving $y_i = w_i - u_i, \forall i$ in U, two boxes are provided to a sampled individual. The first box contains cards labelled $\prime(A \cup B)\prime$ and $\prime(A^C \cap B^C)\prime$ in proportions p_1, $(0 < p_1 < 1)$ and $(1 - p_1)$, respectively, and the second box contains cards marked $\prime(B \cap A^C)\prime$ and $\prime(A \cup B^C)\prime$ in proportions p_2, $(0 < p_2 < 1)$ and $(1 - p_2)$, respectively, and $p_1 \neq \frac{1}{2}$, $p_2 \neq \frac{1}{2}$. The individual i performs inverse Bernoullian trials in each of the two boxes to generate two independent responses, viz. t_{1i} and t_{2i} as given below,

t_{1i} = trial number on which ith respondent gets a match for the first time from1stbox

t_{2i} = trial number on which ith respondent gets a match for the first time from2ndbox.

Then, taking E_R as the RR-based expectation operator and V_R, the RR-based variance operator,

$$r_i = s_i - z_i \quad E_R(r_i) = E_R(s_i) - E_R(z_i) = w_i - u_i = y_i$$

$$s_i = \{t_{1i}(1 - p_1) - 1\}\frac{p_1}{(1 - 2p_1)} z_i = \{t_{2i}(1 - p_2) - 1\}\frac{p_2}{(1 - 2p_2)}.$$

Thus, r_i is an unbiased estimator of y_i with variance,

$$V_R(r_i) = \frac{(1 - p_1)^3 - p_1^3}{(1 - 2p_1)^2} w_i + \frac{p_1^3}{(1 - 2p_1)^2} + \frac{(1 - p_2)^3 - p_2^3}{(1 - 2p_2)^2} u_i + \frac{p_2^3}{(1 - 2p_2)^2},$$

and its estimate as

$$v_i = \frac{(1 - p_1)^3 - p_1^3}{(1 - 2p_1)^2} s_i + \frac{p_1^3}{(1 - 2p_1)^2} + \frac{(1 - p_2)^3 - p_2^3}{(1 - 2p_2)^2} z_i + \frac{p_2^3}{(1 - 2p_2)^2}$$

such that $E_R(v_i) = V_R(r_i)$.

Utilizing Singh and Sedory's (2013) inverse Hypergeometric trials in generating RR, Dihidar (2016) proved that using a sample chosen by any general sampling design, inverse Hypergeometric trials outperform the direct Hypergeometric trials in Kuk's (1990) RR device.

Motivated by Singh and Sedory's (2013) negative Hypergeometric trials in generating RR, Shaw and Chaudhuri (2022) attempt to use the direct and inverse Hypergeometric trials' approach to improve Chaudhuri and Shaw's (2016) revised URL device. The detail estimation procedure is narrated in Sect. 10.3 of this book.

References

Chaudhuri, A. (2011). *Randomized response and indirect questioning techniques in surveys.* Chapman and Hall, CRC Press, Taylor and Francis Group.

Chaudhuri, A., & Dihidar, K. (2014). Generating randomized response by inverse mechanism. *Model Assisted Statistics and Applications, 9*, 343–351.

Chaudhuri, A., & Pal, S. (2002). On certain alternative mean square error estimators in complex survey sampling. *Journal of Statistical Planning and Inference, 104*, 363–375.

Chaudhuri, A., & Shaw, P. (2016). Generating randomized response by inverse Bernoulli an trials in unrelated characteristics model. *Model Assisted Statistics and Applications, 11*, 235–245.

Dihidar, K. (2016). Estimating sensitive population proportion by generating randomized response following direct and inverse hypergeometric distribution. In A. Chaudhuri, T. C. Christofides, & C. R. Rao (Eds.), *Data gathering, analysis and protection of privacy through randomized response techniques: qualitative and quantitative human traits, handbook of statistics,* (Vol. 34, pp. 427–441). North-Holland, Elsevier B.V.

Greenberg, B. G., Kuebler, R. R., Abernathy, J. R., & Horvitz, D. G. (1971). Application of the randomized response technique in obtaining quantitative data. *Journal of the American Statistical Association, 66*(334), 243–250.

Kuk, A. Y. C. (1990). Asking sensitive questions indirectly. *Biometrika, 77*(2), 436–438.

Shaw, P., & Chaudhuri, A. (2022). Further improvements on unrelated characteristic models in randomized response techniques. *Communications in Statistics—Theory and Methods, 51 (2),* 7305–7321.

Singh, S., & Grewal, I. S. (2013). Geometric distribution as a randomization device: Implemented to the Kuk's model. *International Journal of Contemporary Mathematical Sciences, 8*(5), 243–248.

Singh, S., & Sedory, S. A. (2013). A new randomized response device for sensitive characteristics: An application of the negative hypergeometric distribution. *Metron, 71*(1), 3–8.

Walpole, R. E., & Myers, R. H. (1993). *Probability and Statistics for engineers and scientists* (5th ed.). Prentice-Hall.

Warner, S. L. (1965). Randomized response: A survey technique for eliminating evasive answer bias. *Journal of American Statistical Association, 60,* 63–69.

Chapter 9
Other Topics Beyond Chaudhuri (2011) and Chaudhuri and Christofides (2013)

9.1 Introduction

First, Franklin's (1989) RRT is based on SRSWR, but the RRs from sampled persons are gathered by requesting them to draw observations from specified distributions of two different kinds, one to be chosen if the person bears A or from other if he/she bears A^c as in Kuk's (1990). How optional RRs are to be drawn are discussed by others to be selectively covered below. Moment estimators for θ and the proportion of people bearing A are studied by various researchers. But, in addition, sensitivity level estimation of items of enquiry is also studied by various authors. Also, extension to general sampling schemes is implemented by various researchers as are to be briefly illustrated.

9.2 Franklin's (1989) RRT Model

To our utter dissatisfaction, Franklin (1989) considers an SRSWR in n draws. A sampled person i, say, is requested to independently draw $k(> 1)$ times an observation from an explicitly specified probability distribution with a density $g(.)$. if he/she bears A or from a different density $h(.)$ if he/she bears A^c.

Writing θ as the unknown proportion in the community bearing A, then denoting by z_j, the outcome of the jth$(j = 1, \ldots, k)$ draw, the probability density for the outcomes of the $k(> 1)$ draws is

$$\theta \prod_{j=1}^{k} g(z_j) + (1 - \theta) \prod_{j=1}^{k} h(z_j).$$

A. Chaudhuri et al., *Randomized Response Techniques*,
https://doi.org/10.1007/978-981-99-9669-8_9

Using these observations, moment estimation of θ is easily achieved by Franklin (1989) with variance of the estimate $\hat{\theta}$ for θ is easily derived along with an unbiased estimate thereof in view of the simplicity of SRSWR restriction, though there is no history of large-scale survey by SRSWR by any organization ever or anywhere.

9.3   Practical Implementation of Franklin's (1989) RRT by Marcheselli and Barabesi (2006)

Two decks of similar cards in a large number M for both are taken, coloured red and non-red in proportion $p_1 : (1 - p_1)$ in the 1st deck and $p_2 : (1 - p_2)$ in the second, $(0 < p_1 < 1), (0 < p_2 < 1), p_1 \neq p_2$. A sampled person is requested to use the 1st box if he/she bears A and the 2nd box, if A^c and is to draw from the appropriate box randomly m $(1 < m < M)$ cards by SRSWOR method and repeat the same exercise $k(> 1)$ times and report to the interviewer the total number of red cards $(f,$ say) observed in the $k(> 1)$ draws. Obviously,

$$\text{Prob(``} f \text{'' red cards are followed from 1st box)} = \frac{\binom{Mp_1}{f}\binom{M(1 - p_1)}{m - f}}{\binom{M}{m}}$$

and similarly from the 2nd box.

So, $E(f) = mp_1$ for the 1st box and $E(f) = mp_2$ for the 2nd box and $V(f) = cmp_1(1 - p_1)$ and $V(f) = cmp_2(1 - p_2)$ respectively for the two boxes, with $c = \left(\frac{M-m}{M-1}\right)$.

Besides θ, let the additional parameter required to be estimated be the parameter t which is the proportion of persons bearing the stigmatizing attribute A who readily divulge this secret. Chang and Huang (2001) and Huang (2004) consider simultaneous estimation of θ and t.

As a preliminary step, let X_i denote a variable, which is valued 1 if the ith person bears A and divulges this fact and valued 0, otherwise.

Then, $E(X_i) = t\theta$ because X_i is a Bernoulli variable distributed as $b(1, t\theta)$. $\overline{X} = \frac{1}{n}\sum_{i=1}^{n} X_i$ is thus an unbiased estimator of $t\theta$, and hence, it is negatively biased to estimate θ and the less the value of t, the more is its negativity in bias.

Denoting by $\overline{Y}$ the mean of nk values of the numbers of 'red' cards drawn by the n persons over their k repeated drawings of the cards from the 1st box, the unbiased estimator for θ given by Marcheselli and Barabesi (2006) is

$$\hat{\theta}_{MB} = \frac{mp_1\overline{X} + \overline{Y} - mp_2}{mp_1 - mp_2},$$

and

$$\text{Var}\big(\hat{\theta}_{MB}\big) = \frac{\theta(1-\theta)}{n} + \frac{cp_1(1-p_1)(1-t)\theta + cp_2(1-p_2)(1-\theta)}{mkn(p_1 - p_2)^2}.$$

Barabesi and Marcheselli (2006a) have given the following alternative estimator for θ as

$$\hat{\theta}_{BM} = \frac{\big(\overline{F} - mp_2\big)}{m(p_1 - p_2)};$$

here, $\overline{F}$ is the observed mean number of red cards by the n sampled persons in their k repeated drawings from the 1st box.

They also give $\text{Var}\big(\hat{\theta}_{BM}\big) = \frac{\theta(1-\theta)}{n} + \frac{cp_1(1-p_1)\theta + cp_2(1-p_2)(1-\theta)}{mkn(p_1-p_2)^2}$ for which an unbiased estimator is

$$v\big(\hat{\theta}\big) = \frac{\hat{\theta}_{BM}\big(1 - \hat{\theta}_{BM}\big)}{n-1} + \frac{cp_1(1-p_1)\hat{\theta}_{BM} + cp_2(1-p_2)\big(1 - \hat{\theta}_{BM}\big)}{mk(n-1)(p_1 - p_2)^2}.$$

They have also presented Bayes estimate, as preferable ones to the above moment estimators. We omit them to keep our level moderate for this book.

9.4 Barabesi's (2008) Simultaneous Estimation of θ, t

Now let us consider a treatment by Barabesi (2008) of the problem of simultaneous estimation of the proportion θ of people bearing a stigmatizing feature A and the sensitivity level t. They defined as the proportion of people bearing A but divulging directly this 'stigma' as t, the sensitivity level.

They define in addition to the variable y which is 1 for a person bearing A but 0 for one bearing A^c, the new variable x which is valued 1 for a person i who bears A and declares to be so. Then, $\theta = \frac{1}{N}\sum_{i=1}^{N} y_i$, $T = \frac{1}{N}\sum_{i=1}^{N} x_i$ and let $t = \frac{T}{\theta}$. Barabesi (2008) considers a sample s chosen with a general sampling design p for which $\pi_i = \sum_{s \ni i} p(s)$ and $\pi_{ij} = \sum_{\substack{s \ni i, j \\ i \neq j}} p(s)$ which are all positive.

Then, he considers general unbiased estimators for θ and T as $\hat{\theta} = \sum_{i \in s} r_i b_{si}$ and $\hat{T} = \sum_{i \in s} x_i b_{si}$ provided an RRT produces an RR as r_i with $E_R(r_i) = y_i$ along with $V_R(r_i) = \emptyset_i$. But, $t = \frac{T}{\theta}$ being a ratio parameter, Barabesi (2008) noted that a ratio estimator for this 'sensitivity level parameter' will of necessity be biased but he discussed the availability of an approximately unbiased estimator for large samples utilizing Taylor series expansion.

Just now we shall present an alternative concept of 'sensitivity level' introduced by Gupta et al. (2002), consider its classical biased estimation and subsequent unbiased estimation-based on two independent samples taken but invariably the samples are taken as SRSWRs.

9.5 Scrambling RR

Eichhorn and Hayre (1983) presented their concepts of RRTs in quite a different style and approach. They introduced the concept of 'scrambled response' and of a 'scrambling variable'. Letting Y as the variable under study for which the unknown mean μ is needed to be estimated, they introduce a scrambling variable S with known parameters of interest like mean ϑ and variance γ^2. They demand from a sampled respondent in an SRSWR in n draws, a scrambled response $Z = Y\frac{S}{\vartheta}$ without revealing the Y value.

With the observed sample values $z_1, z_2, \ldots, z_n$, an unbiased estimate of $\mu = E(Y)$ is $\bar{z} = \frac{1}{n}\sum_{i=1}^{n} z_i$ taking ϑ as 1.

Using the known value of γ^2, an unbiased estimator for $V(\bar{z})$ is easily derived as $s_{\bar{z}}^2 = \frac{1}{n(n-1)}\sum_{i=1}^{n}(z_i - \bar{z})^2$.

Sat Gupta, Bhisham Gupta and Sarjinder Singh (2002) utilize Eichhorn and Hayre's (1983) innovative scrambling RR approach by their innovation of optional scrambling technique with the following trick.

They ask for the optional scrambled data from selected persons in an SRSWR in n draws as $Z = S^X Y$ with S and Y as in Eichhorn and Hayre (1983) but

$$X = 1 \text{ if the response is scrambled}$$
$$= 0 \text{ if the response is direct.}$$

However, 'for every respondent' $W = \text{Prob}(X = 1) = E(X)$ is taken to be 'the same', by Gupta et al. (2002).

This W is defined to be the 'sensitivity level' for every stigmatizing item of enquiry of course as before $\mu = E(Y)$ being the unknown parameter to be estimated, but $E(S) = \vartheta$ and $Var(S) = \gamma^2$ are known but $W = E(X)$ is unknown and required to be estimated since variables S and T are taken as independent of each other and also of Y.

Note that

$$\begin{aligned}
E(Z) &= E\left(S^X Y\right) \\
&= E\left(S^X Y | X = 1\right)Prob(X = 1) + E\left(S^X Y | X = 0\right)Prob(X = 0) \\
&= E(SY)W + E(Y)(1 - W); E(S) = \vartheta = 1 \\
&= E(Y)W + E(Y)(1 - W) \\
&= W\theta + (1 - W)\theta = \theta.
\end{aligned}$$

So, for an SRSWR in n draws for θ an unbiased estimator is $\hat{\theta} = \frac{1}{n}\sum_{i=1}^{n} z_i$ with easily available formula for $V(\hat{\theta})$, which involves unknown W.

Gupta et al. (2002) easily show that their $\hat{\theta}$ has a smaller variance compared to the corresponding variance of the unbiased estimator for θ obtained by Eichhorn and Hayre because $W < 1$.

For the sensitivity level W, Gupta et al. (2002) obtain only a biased estimator observing $\log Z = X \log S + \log Y$ leading to $E(\log Z) = W E(\log S) + E(\log Y)$.

9.6 Arnab's (2004) Modification

Arnab (2004), however, considers

(1) a varying probability sampling design in general as opposed to SRSWR and
(2) permits the probability of choosing the option to go for a scrambling variable to vary from person to person to provide relief to restrictive approach of Gupta et al. (2002). His scrambling variable is not multiplicative but additive. His RR from a sample selected person i in a sample s chosen by a general design is

$$r_i^* = \delta_i y_i + (1 - \delta_i) r_i; i \in s.$$

Here, $\delta_i = 1$ if i chooses to give a 'DR'

$$= 0 \text{ if } i \text{ chooses to give an 'RR'.}$$

r_i is the RR given by following a prescribed RRT.

However, Arnab (2004) did not consider any concept of a 'sensitivity level'. He, however considered, an alternative to estimator Gupta et al.'s (2002) for W.

Sat Gupta et al. (2010) considered a revised model and used two SRSWRs to provide an alternative estimator for the sensitivity level W.

9.7 Huang's (2010) Approach

Huang (2010) gave us a conclusive procedure to tackle the problem of estimating the sensitivity level. He recommends taking two independent SRSWRs' of sizes n_1 and n_2 and prescribes taking from a sampled person i, the responses as

(i) the correct value Y_i or
(ii) the scrambled value $S_i Y_i \neq T_i$.

Here, S_i and T_i s are independent variate values, also independent of Y_i s with means, variances as $\mu_{S_i} = 1, \mu_{T_i} = \theta_i, i = 1, 2$ for the populations whence the two samples are independently drawn, $\sigma_{S_i}^2 = \gamma^2, \sigma_{T_i}^2 = S_i^2$ which are all known.

The ORR for i is then

$$Z_i = (1 - X_i) Y_i + X_i (S_i Y_i + T_i)$$

with $X_i = 1$ if the response is scrambled

$$= 0 \text{ if it is true}$$

such that $E(X_i) = W \ \forall i \in s$.

Then,

$$E(Z_i) = (1 - W)\theta + W(\theta + \theta), E(Y_i) = \theta.$$

Denoting the observed sample means as $\bar{z}_1$ and $\bar{z}_2$ for the two SRSWRs', $\hat{\theta} = \frac{\theta_1 \bar{z}_1 - \theta_2 \bar{z}_2}{(\theta_1 - \theta_2)}$ and $\hat{W} = \frac{\bar{z}_1 - \bar{z}_2}{\theta_1 - \theta_2}$ taking $\theta_1 \neq \theta_2$ are two unbiased estimators of $E(Y) = \theta$ and $E(X_i) = W$ for both $i = 1, 2$.

Huang (2010) also gave easily the formulae for $\text{Var}(\hat{\theta})$ and $\text{Var}(\hat{W})$ and unbiased estimators thereof.

9.8   Diana and Perri's (2010) Approach

Diana and Perri (2010) illustrated several 'scrambled response' models starting with the one given by Saha (2007) and giving three more by themselves each applied only to an SRSWR of respondents in n draws. They are as follows:

Given by Saha (2007), response is $Z = W(Y + U)$, with Y as the study variable with unknown mean and variance, W and U are scrambling variables, independent of each other and also of Y and with known means, variances $\mu_w, \sigma_w^2, \mu_U, \sigma_U^2$, respectively.

Unbiased estimators of θ, as the sample mean of $z-$ values as $\bar{z}$, $\text{var}(\bar{z})$ and $v(\bar{z})$ for $\text{var}(\bar{z})$ are easily obtained.

Diana and Perri's (2010) models are following:

(1) $Z = Y$ with probability p
      $= W(Y + U)$ with probability $(1 - p)$.
(2) $Z = W[\alpha U + (1 - \alpha)Y]$ with α suitably chosen in $(0, 1)$
(3) $Z = \emptyset(Y + U) + (1 - \emptyset)WU$ with $\emptyset$ in $(0, 1)$ suitably taken.

For $\emptyset = 1$, the scrambling is additive as earlier given by Pollock and Bek (1976).

For each of them, unbiased estimates of θ and variances of the estimates and unbiased variance estimates are easily derived and relative efficiencies are compared.

Various other results of popular interest might be illustrated. But we refrain mostly because we are opposed to SRSWR restriction. An SRSWR is never applied in practice in large-scale sample surveys. The results based on SRSWR are rather easily derived though Chaudhuri (2011) showed how almost every RRT is easy to study based on general sampling schemes. We firmly believe that RRTs for years have been disgracefully ignored by the stalwarts among sampling experts palpably because SRSWR is the principal sampling scheme on which most RRT publications thrive for most of the last 60 years or so. Most of the publications on RRT since 2011–2012 refer to Chaudhuri's (2011) text and yet exclusively deal with SRSWR raising doubts if the authors minded at all the messages in that monograph by Chaudhuri (2011). A practice is evident among contributors to RRT to give references to many publications on RRTs. But a careful study of many of them raises doubts if the authors really perused them with care.

References

Arnab, R. (2004). Optional randomized response techniques for complex designs. *Biometrical Journal, 46*, 114–124.

Barabesi, L. (2008). A design-based randomized response procedure for the estimation of population proportion and sensitivity level. *Journal of Statistical Planning and Inference, 138*, 2398–2408.

Barabesi, L., & Marrcheselli, M. (2006a). A practical implementation and Bayesian estimation in Franklin's randomized response procedure. *Communications in Statistics-Simulation and Computation, 35*, 563–573.

Chang, H. J., & Huang, K. C. (2001). Estimation of proportion and sensitivity of a qualitative character. *Metrika, 53*, 269–280.

Chaudhuri, A. (2011). *Randomized response and indirect questioning techniques in surveys.* CRC Press.

Chaudhuri, A., & Christofides, T. C. (2013). *Indirect questioning in sample surveys.* Springer Verlag.

Diana, G., & Perri, F. P. (2010). New scrambled response models for estimating the mean of a sensitive quantitative character. *Journal of Applied Statistics, 37*(11), 1875–1890.

Eichhorn, B., & Hayre, L. S. (1983). Scrambled randomized response methods for obtaining sensitive quantitative data. *Journal of Statistical Planning and Inference, 7*, 307–316.

Franklin, L. A. (1989). A comparison of estimators for randomized response sampling with continuous distributions from dichotomous population. *Communications in Statistics—Theory and Methods, 18*(2), 489–505.

Gupta, S., Gupta, B., & Singh, S. (2002). Estimation of sensitive level of personal interview survey questions. *Journal of Statistical Planning and Inference, 100*, 239–247.

Gupta, S., Shabbir, J., & Sehra, S. (2010). Mean and sensitivity estimation in optional randomized response models. *Journal of Statistical Pranning and Inference, 140*(10), 2870–2874.

Huang, K. C. (2004). A survey technique for estimating the proportion and sensitivity in a dichotomous finite population. *Statistica Neerlandica, 58*, 75–82.

Huang, K. C. (2010). Unbiased estimators of mean, variance and sensitivity level for quantitative characteristics in finite population sampling. *Metrika, 71*, 341–352.

Kuk, A. Y. (1990). Asking sensitive questions indirectly. *Biometrika, 77*(2), 436–438.

Marcheselli, M., & Barabesi, L. (2006). A generalization of Huang's randomized response procedure for the estimation of population proportion and sensitivity level. *Metron, LXIV, 2*, 145–159.

Pollock, K. H., & Bek, Y. (1976). A comparison of three randomized response models for quantitative data. *Journal of American Statistical Association, 71*, 884–886.

Saha, A. (2007). A simple randomized response technique in complex surveys. *Metron , LXV*, 59–66.

Chapter 10
Topics Covered in the Literature Subsequent to Chaudhuri (2011)

10.1 Introduction

Chaudhuri (2011a) started with Godambe's (1955) homogeneous linear unbiased estimator $t_b = \sum_{i=1}^{N} y_i b_{si} I_{si}$ for $Y = \sum_{i=1}^{N} y_i$ in a DR survey with $I_{si} = 1$ or 0 if $i \in$ s or $i \notin s$, b_{si} free of y_i's in $\underline{Y} = (y_1, \ldots, y_i, \ldots, y_N)$ subject to $\sum_{s \ni i} p(s) b_{si} = 1 \forall i \in U$.

Taking nonzero w_i's $i \in U$,

$$d_{ij} = \sum_{s} p(s)(b_{si} I_{si} - 1)(b_{sj} I_{sj} - 1)$$

$$\beta_i = \sum_{j} d_{ij} w_i, \text{ he notes}$$

$$V_p(t_b) = \sum_{i} y_i^2 C_i + \sum_{i \neq j} y_i y_j C_{ij},$$

$$C_i = \sum_{s} p(s) b_{si}^2 I_{si} - 1, C_{ij} = \sum_{s} p(s) b_{si} b_{sj} I_{si} I_{sj} - 1,$$

and alternatively,

$$V_p(t_b) = -\sum_{i} \sum_{j} w_i w_j d_{ij} \left(\frac{y_i}{w_i} - \frac{y_j}{w_j} \right)^2 + \sum_{i} \frac{y_i^2}{w_i} \beta_i.$$

For particular, $t_H = \sum_{i} \frac{y_i}{\pi_i} I_{si}$,

$$V_p(t_H) = \sum_{i} \sum_{j} (\pi_i \pi_j - \pi_{ij}) \left(\frac{y_i}{\pi_i} - \frac{y_j}{\pi_j} \right)^2$$

$$+ \sum_i \frac{y_i^2 \alpha_i}{\pi_i}; \alpha_i = 1 + \frac{1}{\pi_i} \sum_{j \neq i} \pi_{ij} - \sum_i \pi_i.$$

Then, he quotes unbiased variance estimators

$$v_p(t_b) = \sum y_i^2 C_{si} I_{si} + \sum \sum y_i y_j C_{sij} I_{si} I_{sj}$$

with $\sum p(s) C_{si} I_{si} = C_i$, $\sum p(s) C_{sij} I_{sij} = C_{ij}$ and

$$v_p'(t_b) = -\sum_i \sum_j w_i w_j d_{sij} I_{sij} \left(\frac{y_i}{w_i} - \frac{y_j}{w_j} \right)^2 + \sum_i \frac{y_i^2}{w_i} \beta_i \frac{I_{si}}{\pi_i}$$

and

$$v_p(t_H) = \sum_i \sum_j \left(\frac{\pi_i \pi_j - \pi_{ij}}{\pi_{ij}} \right) \left(\frac{y_i}{\pi_i} - \frac{y_j}{\pi_j} \right)^2 I_{sij} + \sum_i \alpha_i \frac{y_i^2}{\pi_i} I_{si}; I_{sij} = I_{si} I_{sj} \text{ provided}$$

$\pi_{ij} > 0 \forall i \neq j$.

If an appropriate RR survey is executed yielding r_i's with $E_R(r_i) = y_i$, $V_R(r_i)$ with $\hat{V}_i = v_R(r_i)$ satisfying $E_R(v_R(r_i)) = V_R(r_i) = V_i$ then he recommends using $e_b = \sum_i r_i b_{si} I_{si}$ as an unbiased estimator for Y with
$V(e_b) = V_p(t_b) + \sum_i V_i(1 + C_i)$ admitting

$$v(e_b) = \sum_i r_i C_{si} I_{si} + \sum_i \sum_{\neq j} r_i r_j C_{sij} I_{sij} + \sum_i \hat{V}_i \left(\frac{1}{\pi_i} + C_{si} \right) I_{si} \text{ as an unbiased}$$

estimator.

Also $V(e_b) = -\sum_i \sum_{\neq j} w_i w_j d_{ij} \left(\frac{y_i}{w_i} - \frac{y_j}{w_j} \right)^2 + \sum_i \frac{y_i^2}{w_i} \beta_i + \sum_i V_i \left(1 - \frac{\beta_i}{w_i} \right)$

admitting an unbiased estimator $v'(e_b) = -\sum_i \sum_{\neq j} w_i w_j \left(\frac{r_i}{w_i} - \frac{r_j}{w_j} \right)^2 d_{sij} I_{sij} +$

$\sum_i \frac{r_i}{w_i} \frac{\beta_i}{\pi_i} I_{si} + \sum_i \hat{V}_i \frac{I_{si}}{\pi_i}$.

He gave detailed formulae covering optional RR survey data as well in elegant ways.

10.2   Quatember's (2016) Mixture of True and Randomized Responses

This is contrasted with R. Mukerjee's (2016) optional randomized response (ORR) for the same purpose. The former splits a sample s into three disjoint parts.

  (i) s_t of truthful respondents
  (ii) s_u of false respondents and
  (iii) s_m of missing units giving no response about bearing A or A^c.

Mukerjee entertains four categories: n_1 giving DR bearing A, n_2 giving DR admitting A^c, RR stating 'yes' numbering n_3 and n_4 giving RR stating 'no'. Interestingly, Quatember (2016) deals with sampling by general means but Mukerjee (2016) restricts to SRSWR even as late as in 2016 though he refers to the Chaudhuri (2011b). Their analyses in different lines are worth mentioning and inviting perusal but are difficult of covering by summarization.

10.3 Inverse RRTs

Singh and Grewal (2013) gave a pioneering technique of Inverse RRT. Chaudhuri and Dihidar (2014), following them gave the following Inverse versions of Warner's (1965), Kuk's, URL RRTs.

Inverse Warner RRT offering a sampled person i a box of similar cards, an RR is recorded as g_i which is the 'draw number' on which i finds a 'match' for the first time in consecutive draws by SRSWR between his/her real feature A or A^c with the card type A or A^c.

With these inverse Bernoulli trials, one gets

$E_R(g_i) = \frac{y_i}{p} + \frac{1-y_i}{1-p}$ and

$V_R(g_i) = y_i \left(\frac{1-p}{p^2}\right) + (1 - y_i)\left(\frac{p}{(1-p)^2}\right)$, vide Walpole and Myers (1993).

Consequently,

$\varphi_i = \frac{g_i-(1-p)^{-1}}{p^{-1}-(1-p)^{-1}} = \frac{p(1-p)g_i-p}{(1-2p)}$ has $E_R(\varphi_i) = y_i$ and $V_R(\varphi_i) = \frac{p^3}{(1-2p)^2} + \frac{1-p+p^2}{(1-2p)} y_i = V_i$ admitting an unbiased estimator $v_i = \frac{p^3}{(1-2p)^2} + \frac{1-p+p^2}{(1-2p)} \varphi_i$.

On the other hand, following Singh and Grewal (2013) the Inverse Kuk RRT is as follows.

Two boxes of similar red and non-red cards are offered to a sampled person i who on request chooses from the 1st box with cards in proportions $p_1 : (1 - p_1)$ if he/she bears A or if he/she bears A^c is to draw cards by SRSWR from the 2nd box and the RR from i is the number k which is the draw number on which the 1st time a Red card is found.

Then, vide Walpole and Myers (1993) follow

$E_R(k_i) = \frac{y_i}{p_1} + \frac{1-y_i}{p_2}$ leading to $\varphi_i = \frac{\left(k_i - \frac{1}{p_2}\right)p_1p_2}{p_2-p_1}$ and

$V_R(k_i) = \frac{y_i(1-p_1)}{p_1^2} + \frac{(1-y_i)(1-p_2)}{p_2^2}$ and hence

$E_R(\varphi_i) = y_i$ and $V_R(\varphi_i) = \frac{p_1^2 p_2^2 V_R(k_i)}{(p_2-p_1)^2}$ which admits an unbiased estimator

$$v_i = \frac{p_1^2 p_2^2}{(p_2 - p_1)^2}\left[\frac{1 - p_2}{p_2^2} + \frac{(p_2 - p_1)(p_1 + p_2 - p_1p_2)}{p_1^2 p_2^2}\varphi_i\right].$$

With these materials at hand it is possible to draw a sample following suitable schemes like SRSWR, SRSWOR, PPSWR, PPSWOR, RHC or IPPS and employing appropriate estimators based on RR data by the Revised—Warner or Revised—Kuk RRTs. Chaudhuri and Dihidar (2014) examined whether it is worthy enough to adopt these two revised schemes rather than the original Warner and original Kuk RRTs. Theoretically, no conclusion is possible about advantage or disadvantage in going for

Singh-Grewal's alternative. So they resorted to simulation studies, and they found in practice it might be marginally advantageous often to go for Singh- Grewal's inverse approach.

Starting with RRTs due to Warner (1965) and Kuk (1990), there was no difficulty in switching over to Singh-Grewal versions of them. But Purnima Shaw (2021) found URL RRTs due to Simmons and his colleagues are not amenable to do this switch over. So, she revised URL of Simmons and could apply Singh-Grewal technique to the revised URL, vide Chaudhuri and Shaw (2016).

To apply Singh-Grewal's (2013) amendment to URL by Simmons, we may, vide Purnima Shaw's Ph.D. Thesis (2021), define the following:

$$w_i = 1 \text{ if } i \text{ bears } (A \cup B).$$
$$= 0 \text{ if } i \text{ bears } (A \cup B)^c = A^c \cap B^c.$$
$$u_i = 1 \text{ if } i \text{ bears } B \cap A^c.$$
$$= 0 \text{ if } i \text{ bears } (B \cap A^c)^c = A \cup B^c.$$

Then, $y_i = w_i - u_i \forall i \in U$.

As it is impossible to have $w_i = 0$ and $u_i = 1$, so y_i equals either 1 or 0 as it must.

Let us have two boxes with one having cards marked $A \cup B$ in proportion $p_1 (0 < p_1 < 1)$ and $A^c \cap B^c$ in proportion $(1 - p_1)$. The second box is to carry cards marked $B \cap A^c$ and $A \cup B^c$ in proportions $p_2 : (1 - p_2)$, $p_1 \neq p_2, (0 < p_2 < 1)$; also $p_1 \neq \frac{1}{2} \neq p_2$.

A sampled person i is to independently draw a card from each box and report.

$I_i = 1$ if finds a 'match' in 1st box for his/her feature and card type.

$= 0$, else.

and

$J_i = 1$ if 'match' in 2nd box.

$= 0$ if 'no match' in 2nd box.

Then, $E_R(I_i) = p_1 w_i + (1 - p_1)(1 - w_i) = (1 - p_1) + (2p_1 - 1)w_i$.

and so, $s_i = \frac{I_i - (1 - p_1)}{2p_1 - 1}$ has $E_R(s_i) = w_i$.

Likewise, $E_R(J_i) = (1 - p_2) + (2p_2 - 1)u_i$ and

$z_i = \frac{J_i - (1 - p_2)}{2p_2 - 1}$ has $E_R(z_i) = u_i$.

So, $r_i = s_i - z_i$ has $E_R(r_i) = y_i$.

Since $I_i^2 = I_i$, $J_i^2 = J_i$, $w_i^2 = w_i$, $u_i^2 = u_i$,

$V_R(s_i) = \frac{p_1(1-p_1)}{(2p_1-1)^2}$ and

$V_R(z_i) = \frac{p_2(1-p_2)}{(2p_2-1)^2}$.

So, $V_R(r_i) = \frac{p_1(1-p_1)}{(2p_1-1)^2} + \frac{p_2(1-p_2)}{(2p_2-1)^2} = V_i.$, say.

Since V_i is known, it needs no estimation.

With this revised URL, amended RRT following Singh and Grewal (2013) easily applies.

Letting

t_{1i} = the trial number at which the 1st 'match' comes from the 1st box.

and t_{2i} = the trial number on which one gets the 1st 'match' in the 2nd box.

Now, Walpole and Myers (1993) give

$$E_R(t_{1i}) = \frac{w_i}{p_1} + \frac{1 - w_i}{1 - p_1}$$

and

$$V_R(t_{1i}) = \frac{w_i(1 - p_1)}{p_1^2} + \frac{(1 - w_i)p_1}{(1 - p_1)^2}$$

$E_R(t_{2i})$ and $V_R(t_{2i})$ are similarly worked out. Then,

$$s_i = \{t_{1i}(1 - p_1) - 1\}\frac{p_1}{(1 - 2p_1)}$$

and

$$z_i = \{t_{2i}(1 - p_2) - 1\}\frac{p_2}{(1 - 2p_2)}, r_i = s_i - z_i.$$

$$E_R(s_i) = w_i, E_R(z_i) = u_i, \text{ and } E_R(r_i) = y_i$$

with a little algebra one may find $V_R(r_i)$ in easy terms of p_1, p_2, w_i and u_i.

This $V_R(r_i)$ admits an unbiased estimator $r_i(r_i - 1) = v_i$, say.

Then, unbiased estimation of θ and unbiased estimation of variance of this estimator are easy to accomplish. So, revised URL and Singh-Grewal version of that may be routinely accomplished with these antecedents.

Next we turn to Purnima Shaw's celebrated works on Direct and Inverse hypergeometric versions of the above Bernoulli aspect of here transformed URL RRT.

Let us take two boxes. In the 1st box, let there be N_1 cards of which $R_1\left(< N_1, \text{ but } R_1 \neq \frac{N_1}{2}\right)$ are labelled '$A \cup B$' each and the rest $N_1 - R_1$ labelled $A^c \cap B^c$ each. The second box, suppose has N_2 cards of which $R_2\left(< N_2, \text{ but } R_2 \neq \frac{N_2}{2}\right)$ are labelled $B \cap A^c$ each and the rest $R_2 - N_2$ labelled $A \cup B^c$ each.

From both the boxes, a sampled person i is requested to independently select at random without replacement $k(\leq N_1, N_2)$ cards and report the numbers f_{1i} for the 1st box and f_{2i} for the 2nd box the numbers respectively of the first 'matches' in the person's true feature versus the card type found.

Then, follow (recalling w_i and u_i).

$E_R(f_{1i}) = k\left\{w_i\frac{R_1}{N_1} + (1 - w_i)\frac{N_1 - R_1}{N_1}\right\}$ and

$V_R(f_{1i}) = k\left(\frac{N_1 - R_1}{N_1}\right)\frac{R_1}{N_1}\left(\frac{N_1 - k}{N_1 - 1}\right) = V_{1i}$, say,

which is known and need not be estimated.

Also, $r_{1i} = \frac{\frac{f_{1i}w_1}{k} - (N_1 - R_1)}{2R_1 - N_1}$ has $E_R(r_{1i}) = w_i$.

So, r_{1i} is an unbiased estimator of w_i with known variance

$$V_R(r_{1i}) = \frac{N_1^2}{k^2(2R_1 - N_1)^2} V_R(f_{1i}).$$

Likewise

$$E_R(f_{2i}) = k\left\{ u_i \frac{R_2}{N_2} + (1 - u_i)\left(\frac{N_2 - R_2}{N_2}\right)\right\}$$

$$V_R(f_{2i}) = k\left(\frac{N_2 - R_2}{N_2}\right)\frac{R_2}{N_2}\left(\frac{N_2 - k}{N_2 - 1}\right).$$

Then, $r_{2i} = \frac{\frac{f_2 w_2}{k} - (N_2 - R_2)}{2R_2 - N_2}$ has $E_R(r_{2i}) = u_i$ and $V_R(r_{2i}) = \frac{N_2^2}{k^2(2R_2 - N_2)^2} V_R(f_{2i})$.

Then, $r_i = r_{1i} - r_{2i}$ has $E_R(r_i) = y_i$ and $V_R(r_i) = V_R(r_{1i}) + V_R(r_{2i}) = V_i$, say, is a known quantity.

So, with these antecedents it is easy to unbiasedly estimate $\theta = \frac{1}{N}\sum y_i$ along with an unbiased estimate of its variance considering suitable sampling schemes and appropriate estimators for θ.

Next, honouring Singh-Grewal (2013) approach let us consider the inverse hypergeometric trials-based RRT of Simmons's URL revised by Purnima Shaw.

Let a sampled person i be requested to independently from two boxes draw by SRSWOR and to report the following two results, viz.

g_{1i} = trial number at which $T_1(T_1 \leq R_1, T_1 \leq N_1 - R_1)$ matches are achieved for the 1st box and

g_{2i} = trial number at which $T_2(T_2 \leq R_2, T_2 \leq N_2 - R_2)$ matches are achieved for the 2nd box.

Then, $E_R(g_{1i}) = T_1\left\{ w_i \frac{N_1+1}{R_1+1} + (1 - w_i)\frac{N_1+1}{N_1-R_1+1}\right\}$.

Letting $r_{1i} = \frac{g_{1i} - T_1\left(\frac{N_1+1}{N_1-R_1+1}\right)}{T_1\left\{\frac{N_1+1}{R_1+1} - \frac{N_1+1}{N_1-R_1+1}\right\}}$, one has $E_R(r_{1i}) = w_i$,

$V_R(r_{1i}) = \frac{V_R(g_{1i})}{T_1^2\left\{\frac{N_1+1}{R_1+1} - \frac{N_1+1}{N_1-R_1+1}\right\}^2}$ which admits an unbiased estimator

$v_{1i} = \frac{a r_{1i} + b}{T_1^2\left\{\frac{N_1+1}{R_1+1} - \frac{N_1+1}{N_1-R_1+1}\right\}^2}$ with $a = T_1(N_1 + 1)\left\{\frac{(N_1-R_1)(R_1+1-T_1)}{(R_1+1)^2(R_1+2)} - \frac{R_1(N_1-R_1+1-T_1)}{(N_1-R_1+1)^2(R_1+2)}\right\}$

and $b = T_1\frac{R_1(N_1-R_1+1-T_1)(N_1+1)}{(N_1-R_1+1)^2(R_1+2)}$.

Similar formulae follow starting with g_{2i} with $E_R(g_{2i}) = u_i$, $V_R(r_{2i})$ admitting an unbiased estimator v_{2i} so that $r_i = r_{1i} - r_{2i}$ has

$E_R(r_i) = w_i - u_i = y_i$ and

$$V_R(r_i) = V_R(r_{1i}) + V_R(r_{2i})$$

with v_{1i} and v_{2i} as unbiased estimators for $V_R(r_{1i})$ and $V_R(r_{2i})$ so that $v_i = v_{1i} + v_{2i}$ as an unbiased estimator for $V_R(r_i)$.

Then, adopting suitable sampling schemes and unbiased estimators for θ using RR survey data at hand simulation studies are performed. Then, simulation results based on revised URL in its version and the Singh-Grewal version thereof suitable comparisons are actually studied in the literature to judge if Singh-Grewal may yield higher efficiency.

Dihidar (2016), Chaudhuri and Shaw (2016) and Shaw and Chaudhuri (2022) are three references relevant here.

10.4 Three Ensuing Messages

1. In a paper possibly to be published soon a scrambled response on seemingly stigmatizing issue the following models are considered.

   (i) With Y as the study variable with unknown mean, variance θ, σ^2 and W as a scrambling variable with known mean, variance μ_w, σ_w^2, overall-known model of Gjestvang and Sarjinder Singh (2009) treats a scrambled response from a person i in an SRSWR in n draws as

   $z_i = Y_i + \alpha W_i$ with probability $p = \frac{\beta}{\alpha+\beta}$.
   $= Y_i - \beta W_i$ with probability $1 - p = \frac{\alpha}{\alpha+\beta}$.
   with α, β as known positive numbers.

   For this unbiased estimator for θ is $\hat{\theta}_{GS} = \frac{1}{n} \sum_{i=1}^{n} z_i$ with

   $$\text{var}(\hat{\theta}_{GS}) = \frac{1}{n}\left[\alpha\beta(\sigma_w^2 + \mu_w^2) + \sigma^2\right].$$

   (ii) Narjis and Shabbir (2021) models:

   $z_i = Y_i - \beta W_i$ with probability $p_1 = \frac{\alpha}{\alpha+\beta+\gamma}$.
   $= Y_i + \alpha W_i$ with probability $p_2 = \frac{\beta}{\alpha+\beta+\gamma}$.
   $= Y_i$ with probability $p_3 = \frac{\gamma}{\alpha+\beta+\gamma}$.
   with γ as also a given positive number.

   Unbiased estimator θ from an SRSWR (n) is $\frac{1}{n} \sum_{i=1}^{n} z_i = \bar{z}$ with

   $$\text{Var}(\bar{z}) = \frac{1}{n}\left[\frac{\alpha\beta(\alpha - \beta)(\sigma_w^2 + \mu_w^2)}{\alpha + \beta + \gamma} + \sigma^2\right].$$

   (iii) For the model proposed by the anonymous forthcoming author(s),

   $z_i = \frac{\lambda+1}{2}Y_i - \frac{\beta}{2}W_i$ with probability $p_1 = \frac{\alpha}{\lambda(\alpha+\beta)}$.
   $= \frac{\lambda+1}{2}Y_i + \frac{\alpha}{2}W_i$ with probability $p_2 = \frac{\beta}{\lambda(\alpha+\beta)}$.
   $= \frac{Y_i}{2}$ with probability $p_3 = 1 - \frac{1}{\lambda}$ (obviously $\lambda > 1$).

   An unbiased estimator for θ based on an SRSWR (n) is $\bar{z} = \frac{1}{n} \sum_{i=1}^{n} z_i$

   and $Var(\bar{z})$ meticulously derived. Measures of privacy protections and efficiency comparison are zealously accomplished.

Two reasons, why we present this work are (1) the paper is very new and (2) SRSWR is still duly respected in RR context.

1. In a promising forthcoming paper by three well-known authors Franklin (1989) RRT is being revisited with a novelty of correlated scrambling variables all in the old bottle of SRSWR.

2. Finally, let us mention the paper "Respondent-specific Randomized Response Technique" to estimate sensitive proportion jointly by all the three authors of this monograph. This is to appear in 'Statistics in Transition-New Series' in the final or 4th issue of this in 2023. This is referred to in the bibliography of the present chapter. This considers RR data collection by a varying probability sampling scheme to estimate the proportion of people in a community bearing a sensitive characteristic A. But the device offers to a sampled person two boxes with five types of cards identified by five alternatives 'I possess A', 'I possess A^c', 'I possess the innocuous feature B', 'yes' and 'no' in varying proportions in both the boxes. This is done to offer a respondent enough assurance about maintenance of secrecy protected to a highly desirable extent. The authors give two separate models for generation of RRs. The analysis is done in an enhanced complexity but the authors believe that privacy is well protected with less loss in efficient estimation suffered in classical RRTs. Measures of privacy protection are well discussed. Enough simulation studies are presented quoting live data from Chaudhuri et al. (2009).

References

Chaudhuri, A. (2011a). *Randomized response and indirect questioning techniques in surveys*. CRC Press.

Chaudhuri, A. (2011b). Unequal probability sampling: analyzing optional randomized response on qualitative and quantitative variables bearing social stigma. In *International Statistical Institute Proceedings of the 58th World Statistics Congress*, (pp. 1939–1947). Dublin.

Chaudhuri, A., & Dihidar, K. (2014). Generating randomized response by inverse mechanism. *Model Assisted Stat Appl, 9*, 343–351.

Chaudhuri, A., & Shaw, P. (2016). Generating randomized response by inverse bernoulli trials in unrelated characteristics model. *Model Assisted Statistics and Applications, 11*, 235–245.

Chaudhuri, A., Christofides, T. C., & Saha, A. (2009). Protection of privacy in efficient application of randomized response techniques. *Stat Methods and Appl, 18*, 389–418.

Dihidar, K. (2016). Estimating sensitive population proportion by generating randomized response following direct and inverse hypergeometric distribution. In A. Chaudhuri, T. C. Christofides, & C. R. Rao (Eds.), *Data gathering, analysis and protection of privacy through randomized response techniques: Qualitative and quantitative human traits, handbook of statistics, 34* (pp. 427–441). Elsevier.

Franklin, L. A. (1989). A comparision of estimators for randomized response sampling with continuous distributions from dichotomous population. *Communications in Statistics - Theory and Methods, 18*(2), 489–505.

Gjestvang, C. R., & Singh, S. (2009). An improved randomized response model: Estimation of mean. *Journal of Applied Statistics, 36*(12), 1361–1367.

Godambe, V. P. (1955). A unified theory of sampling from finite populations. *Journal of Royal Statistical Society B, 17*, 269–278.

Kuk, A. Y. (1990). Asking sensitive questions indirectly. *Biometrika, 77*(2), 436–438.

Mukerjee, R. (2016). Optional Randomized Response Revisited. In A. Chaudhuri, T. C. Christofides, & C. R. Rao (Eds.), *Data gathering, analysis and protection of privacy through randomized response techniques: qualitative and quantitative human traits, handbook of statistics, 34* (pp. 331–340). Elsevier.

Patra, D., Pal, S., & Chaudhuri, A. (2023). Respondent-specific randomized response technique to estimate sensitive proportion. *Statistics in Transition, 24*(4), 139–156.

Quatember, A. (2016). A mixture of true and randomized responses in the estimation of the number of people having a certain attribute. In A. Chaudhuri, T. C. Christofides, & C. R. Rao (Eds.), *data gathering, analysis and protection of privacy through randomized response techniques: Qualitative and quantitative human traits, handbook of statistics* (Vol. 34, pp. 105–117). Elsevier.

Shaw, P., & Chaudhuri, A. (2022). Further improvements on unrelated characteristic models in randomized response techniques. *Communications in Statistics-Theory and Methods, 21*, 7305–7321.

Shaw, P. (2021). Innovative contributions to estimation techniques in finite population framework *(Unpublished Thesis)*. West Bengal State University.

Singh, S., & Grewal, I. S. (2013). Geometric distribution as a randomized device: Implemented to the Kuk's model. *International Journal of Contemporary Mathematical Sciences, 8*(5), 243–248.

Walpole, R. E., & Myers, R. H. (1993). *Probability and Statistics for engineers and scientists* (5th ed.). Prentice-Hall.

Warner, S. L. (1965). Randomized response: A survey technique for eliminating evasive answer bias. *Journal of American Statistical Association, 60*, 63–69.

Chapter 11
Alternatives to RRT in Studying Sensitive and Stigmatizing Issues

11.1 Introduction

For surveys involving sensitive questions, randomized response techniques (RRTs) and other indirect questions are helpful in obtaining survey responses while maintaining the privacy of the respondents. The RRT is controlled by interviewers, which makes it difficult to convince the respondents that their privacy is protected by randomization. Alternatives to RRTs are also coming up in recent years. Other indirect questioning includes three-card method, several non-randomized methods, the item count technique, nominative technique to estimate finite population proportion θ of a sensitive characteristic A.

The three-card (Droitcour et al., 2001) method was originally designed to survey foreign-born persons about their immigration status (including illegal or undocumented status); however, it can be applied in a variety of other sensitive question areas. It is a relatively new survey approach designed to estimate both the size and characteristics of the resident illegal population.

Tian et al. (2007), Yu et al. (2008), and Tan et al. (2009) have developed an interesting "non-randomized response approach". The non-randomized response approach offers an attractive alternative to the randomized response approach. In non-randomized models, no randomization device is needed and the method of surveying with negative questions is employed. The questions are asked in a negative way so that a respondent can provide one of many possible answers. These approaches protect respondent privacy vis-à-vis the interviewer and others. Such techniques promise that no one would ever be able to discover a respondent's status with respect to the sensitive question. Yu et al. (2008) illustrated only "triangular and crosswise models" of data gathering Devices. Christofides (2009), essentially with the preceding situation illustrated, applied his non-randomized response (NRR) technique (NRRT) permitting only an SRSWR, in a different way narrated. Chaudhuri (2012) narrated how NRR schemes in vogue may be used in unbiased estimation even when a sample may be selected following a general scheme.

A. Chaudhuri et al., *Randomized Response Techniques*,
https://doi.org/10.1007/978-981-99-9669-8_11

Takahasi and Sakasegawa (1977) should be noted as the pioneers in advocating NRR to estimate sensitive proportion θ. Pal (2007a) extended Takahasi and Sakasegawa's (1977) NRRT to general sampling situations. Another indirect questioning technique is known as the "nominative technique" (NT) in which a sampled person is requested to report about one or more acquaintances as to whether they bear a sensitive attribute but he/she is not required to reveal whether he/she personally belongs to the stigmatizing group.

11.2   Non-randomized Techniques

Contrasting with well-publicized randomized response (RR) techniques (RRT) claimed to be useful in estimating the proportion of people bearing a sensitive characteristic in a given community; recently non-randomized response (NRR) techniques (NRRT) are emerging. With most early RRTs, the NRRTs are also applied to samples selected by simple random sampling (SRS) with replacement (SRSWR) scheme alone.

Originally, two non-randomized response models were proposed by Yu et al., namely the triangular and crosswise models. Unlike the RR models, the NRR models use an independent non-sensitive question (e.g. season of birth) in the questionnaire to obtain indirectly a respondent's answer to a sensitive question. Obviously, the NRR designs reduce costs as needed in RR surveys. The different NRR models are narrated in this section.

11.2.1   Triangular Model

Consider a finite population $U = (1, 2, \ldots, i, \ldots, N)$ of units labelled $1, 2, \ldots, i, \ldots, N$, each of which is distinctly identifiable. Let x and y be two real variables with values y_i and x_i, with i in U. Let y_i take value 1 if the ith$(i = 1, 2, \ldots, N)$ individual in the population U bears a sensitive characteristic A (e.g. drug taking) and value 0 if individual i bears A^C. Similarly x_i takes value 1 if the ith$(i = 1, 2, \ldots, N)$ individual in the population U bears a non-sensitive characteristic X and value 0 if individual i bears X^C. For example, we may define $x_i = 1$ if the respondent was born between August and December, and $x_i = 0$ otherwise. Our aim is to estimate the proportion of people bearing the characteristic A. The interviewer should select a suitable X so that the proportion $p = P(x = 1)$ can be estimated easily. The proportion p may be assumed to be known.

Under the triangular model, a non-sensitive question with known population prevalence is introduced. The question and answer options are placed in a 2×2 contingency table where two quadrants relate to the innocuous questions with known population prevalence while the other two quadrants represent the binomial response options to the sensitive question. Such a design encourages respondents not only to participate in the survey, but also to provide their truthful responses.

For a face-to-face interview, the interviewer may use the format of Table 11.1.

A simple random sample of size n is drawn from the population U. The respondent is asked to use the format at the left side of Table 11.1.

The respondent will put a tick in the circle (i.e. $\{y = 0, x = 0\}$) if he/she belongs to this circle or put a tick in the upper square (i.e. $\{y = 0, x = 1\}$) if he/she belongs to one of the three squares. So he/she will put a tick in either the open circle or in the triangle formed by the three solid dots in Table 1 according to his or her truthful answer. Note that both $\{y = 0, x = 0\}$ and $\{y = 0, x = 1\}$ are non-sensitive, and the sensitive class $\{y = 1\}$ is mixed with another non-sensitive subclass $\{y = 0, x = 1\}$.

Here, $\{y = 0, x = 0\}$ represents a non-sensitive subclass. Similarly, $\{y = 1\} \cup \{y = 0, x = 1\}$ is also a sensitive subclass. This design encourages respondents to provide their truthful responses. Yu et al. (2008) called this a triangular model.

If $x = 1$ represents that a respondent was born between July and December and $p = \frac{1}{2}$, then we can reformulate the triangular model into a non-sensitive question of the following form:

If a respondent does not belong to sensitive group, (i.e. according to your actual birthday) put a tick in the following circle, i.e. O) or square (i.e. ■). Otherwise, please put a tick in the following square regardless of his/her actual birthday.

(a) I was born in the first half of a year, check here o
(b) I was born in the second half of a year, check here ■.

This non-randomized design encourages cooperation from the respondents, and their sensitive characteristics will not be exposed to others.

One immediate advantage of the triangular model is its robustness to non-response in the sense that it allows such non-response. For example, for n respondents, we observed n_1 ticks in the circle, n_2 ticks in the triangle, and n_3 non-responses. In other words, we observed n_1 ticks in the circle and $n_2 + n_3$ ticks in the triangle under the assumption that a respondent is always willing to answer the question if he or she belongs to non-sensitive group.

For the triangular design, we define a "hidden" variate Y^{HT} as follows:

$$Y^{HT} = \begin{cases} 1 \text{ with probability } \theta + (1 - \theta)p \text{ if a tick is put in the traingle} \\ 0 \text{ with probability } (1 - \theta)(1 - p) \text{ if a tick is put in the circle.} \end{cases}$$

Table 11.1 Triangular model and the corresponding cell probabilities

Categories	$x = 0$	$x = 1$	Categories	$x = 0$	$x = 1$	Marginal
$y = 0$	O	■	$y = 0$	$(1 - \theta)(1 - p)$	$(1 - \theta)p$	$1 - \theta$
$y = 1$	■	■	$y = 1$	$\theta(1 - p)$	θp	θ
			Marginal	$1 - p$	p	1

Please put a tick in the circle if you belong to this circle or put a tick in the upper square if you belong to one of the three squares

So,

$$Y_i^{HT} = \begin{cases} 1 & \text{if the } i\text{th responsdent puts a tick in the triangle} \\ 0 & \text{else} \end{cases}.$$

The likelihood function for θ is

$$L(\theta|Y) = \prod_{i=1}^{n} [\theta + (1-\theta)p]^{y_i^{HT}} [(1-\theta)(1-p)]^{1-y_i^{HT}}.$$

The MLE for θ is $\hat{\theta}_T = \frac{\overline{Y}_T - p}{1-p}$, where $\overline{Y}_T = \frac{1}{n}\sum_1^n Y_i^{HT}$.

Its variance is given by $\text{Var}\left(\hat{\theta}_T\right) = \text{Var}\left(\hat{\theta}_D\right) + \frac{p(1-\theta)}{n(1-p)}$, where $\text{Var}\left(\hat{\theta}_D\right) = \frac{\theta(1-\theta)}{n}$, the variance for the direct questioning on sensitive character.

For any fixed θ,

$n \, \text{Var}(\hat{\theta}_T) = (1-\theta)\left[\theta + \frac{p}{1-p}\right]$ is an increasing function of p ($0 < p < 1$) and slowly approaches infinity as $p \to 1$.

11.2.2   Crosswise Models

Yu et al. (2008) introduced a so-called crosswise model, which can be viewed as a non-randomized version of the original Warner model. The interviewer may design a questionnaire in the format as shown on the left-hand side of Table 11.2 and asks each interviewee to put a tick in the upper circle (i.e. $\{y = 0, x = 0\}$) if he/she belongs to one of the two circles or put a tick in the upper square (i.e. $\{y = 0, x = 1\}$) if he/she belongs to one of the two squares. Note that both $\{y = 0, x = 0\}$ and $\{y = 0, x = 1\}$ are non-sensitive. Thus, an interviewee who belongs to the sensitive class (i.e. $\{y = 0\}$ is not being exposed if a tick is put in the upper circle/square. The corresponding cell probabilities are listed at the right-hand side of Table 11.2. Yu et al. (2008) called this the crosswise model.

The respondent will put a tick in the upper circle if he/she belongs to one of the two circles or put a tick in the upper square if he/she belongs to one of the two squares.

Table 11.2 Crosswise model and the corresponding cell probabilities

Categories	$x = 0$	$x = 1$	Categories	$x = 0$	$x = 1$	Marginal
$y = 0$	O	■	$y = 0$	$(1-\theta)(1-p)$	$(1-\theta)p$	$1-\theta$
$y = 1$	■	■	$y = 1$	$\theta(1-p)$	θp	θ
			Marginal	$1-p$	p	1

Suppose there are a total of n ticks (corresponding to n^* respondents in the survey without non-response) with n^* ticks being put in the upper circle and $n - n^*$ ticks being put in the upper square.

The probability of putting a tick in the upper circle is $(1 - \theta)(1 - p) + \theta p$.

Let $\lambda = (1 - \theta)(1 - p) + \theta p$.

Then, $\theta = \frac{p-1+\lambda}{2p-1}$, where $p \neq \frac{1}{2}$ is known.

The likelihood $L \approx \lambda^{n^*}(1 - \lambda)^{n-n^*}$.

The MLE of λ is given by $\hat{\lambda} = \frac{n^*}{n}$.

So the MLE of θ is $\hat{\theta}_C = \frac{p-1+\hat{\lambda}}{2p-1}$, where $p \neq \frac{1}{2}$ is known provided that $0 < \hat{\theta}_C < 1$.

If $\hat{\theta}_C \langle 0 or \rangle 1$, then EM algorithm may be used. The details are narrated in Chap. 2 of the monograph of Tian and Tang (2014). Here, $E\left(\hat{\theta}_C\right) = \theta$.

The variance is $Var\left(\hat{\theta}_C\right) = \frac{\lambda(1-\lambda)}{n(2p-1)^2} = \frac{\theta(1-\theta)}{n} + \frac{p(1-p)}{n(2p-1)^2}$.

The MLE and the corresponding variance are identical to the original estimate of θ and its variance in Warner's model. Hence, Yu et al. (2008) considered the crosswise model as a non-randomized version of the Warner model. It can be noted that in non-randomized crosswise model no randomization device is required.

11.2.3 Parallel Model

The crosswise and triangular models require that one category is non-sensitive. Thus, they cannot be applied to a situation where two categories are sensitive like income, disloyal or loyal to his/her boss and so on. Tian and Tang (2014) proposed a non-randomized version of the unrelated question model, called the parallel model with a wider application range.

Let $\{y = 1\}$ denote the class of people who possess a sensitive characteristic and $\{y = 0\}$ denote the complementary class. The objective is to estimate the proportion $\theta = \Pr(y = 1)$.

Suppose that X and W are two non-sensitive dichotomous variates. Let x_i take value 1 if the $i^{th}(i = 1, 2, \ldots, N)$ individual in the population U bears a non-sensitive characteristic X and value 0 if individual i bears X^C.

Similarly, the other variable w_i takes value 1 if the $i^{th}(i = 1, 2, \ldots, N)$ individual in the population U bears a non-sensitive characteristic W and value 0 if individual i bears W^C.

Suppose that X and W are two non-sensitive dichotomous variates, and A, X and W are mutually independent with known $q = \Pr(w = 1)$ and $p = \Pr(x = 1)$. For example, we may define $w = 1$ if the respondent's birthday is in the second half of a month and $w = 0$ otherwise. Similarly, $x = 1$ if the respondent was born between July and December and $x = 0$ otherwise. Hence, $q \approx .5$ and $p \approx .5$ can be assumed.

Interviewers may design a questionnaire in the format as shown at the left-hand side of Table 11.3 and ask each interviewee to connect the two circles by a straight line if he/she belongs to one of the two circles or connect the two squares by a straight line if he/she belongs to one of the two squares. Note that all $\{x = 0\}$, $\{x = 1\}$, $\{w = 0\}$ and $\{w = 1\}$ and $\{y = 0\}$ are non-sensitive classes; thus, $\{w = 1, x = 0\} \cup \{y = 1, x = 1\}$ is also a non-sensitive subclass.

The corresponding cell probabilities are displayed at the right-hand side of Table 11.3.

The respondents connect the two circles by a straight line if they belong to one of the two circles or connect the two squares by a straight line if they belong to one of the two squares.

Suppose there are a total of n straight lines (corresponding to n respondents in the survey) with $n - n^p$ lines connecting the two circles and n^p lines connecting the two squares (see Table 11.3).

The probability of drawing a line to connect the two squares is then $q(1 - p) + \theta(p)$. Defining it by a new parameter $\lambda = q(1 - p) + \theta(p)$, which is identical to unrelated question RR model, we can proceed as illustrated in crosswise model (vide 11.2.2).

The MLE for θ is $\hat{\theta}_P = \frac{\hat{\lambda} - q(1-p)}{p}$.

In general, the interviewers do not know the information of birth date of a respondent.

Hence, the interviewers receive the response (e.g. 'yes' or 'no') from a respondent without the knowledge of which question is being answered by the respondent.

Unfortunately, both randomized and non-randomized response models could underestimate the proportion of subjects with the sensitive characteristic as some respondents do not believe that these techniques can protect their privacy. Taking the non-compliance into consideration, Wu and Tang (2016) introduced new non-randomized response techniques in which no covariate is required.

The "three-card method" (TCM) is also an alternative indirect questioning technique.

Droitcour et al. (2001) gave this method using three independent samples, in each of which, a person is given three separate boxes with three cards bearing identification about characteristics, namely the sensitive one as A and the innocuous ones like B, C and D, say. Each sampled person will choose one of the three boxes and announce the box number only. But it was restricted to SRSWR only. Chaudhuri and Christofides (2008) extended it to general sampling schemes. Assume that we have three boxes, Box 1, Box 2 and Box 3. Let the composition of cards in three boxes be as follows:

Table 11.3 Parallel model and the corresponding cell probabilities

Categories	$x = 0$	$x = 1$	Categories	$x = 0$	$x = 1$	Marginal
$w = 0$	O	■	$w = 0$	$(1 - q)(1 - p)$		$1 - q$
$w = 1$	■		$w = 1$	$q(1 - p)$		q
$y = 0$		O	$y = 0$		$(1 - \theta)p$	$1 - \theta$
$y = 1$		■	$y = 1$		θp	θ
			Marginal	$1 - p$	p	1

The first sample is presented with three boxes each one including statements as follows:

Box 1: I belong to B.
Box 2: I belong to C or D or A.
Box 3: I belong to some other group not in Box 1 or Box 2.

The second sample is presented with the three boxes as follows:
Box 1: I belong to C.
Box 2: I belong to B or D or A.
Box 3: I belong to some other group not in Box 1 or Box 2.

The third sample is presented with the following:
Box 1: I belong to D.
Box 2: I belong to B or C or A.
Box 3: I belong to some other group not in Box 1 or Box 2.

The groups must be mutually exclusive. Then,

$$\hat{\theta}_{1A} = \hat{\theta}_{1ACD} - \hat{\theta}_C - \hat{\theta}_D$$

is the estimator of the population proportion of people having the sensitive characteristic where θ_B, θ_C and θ_D are the population proportions of persons bearing the innocuous character B, C and D. Let, $\hat{\theta}_B, \hat{\theta}_C$ and $\hat{\theta}_D$ be the estimates of θ_B, θ_C and θ_D based on sample 1, sample 2 and sample 3, respectively.

11.2.4 General Sampling by Three Non-randomized Response (NRR) Schemes

Chaudhuri (2012) showed how the NRR schemes may be used in unbiased estimation even when a sample may be selected following a general sampling scheme.

The modifications for triangular model are as follows:

Let $U = (1, 2, \ldots, i, \ldots, N)$ denote a finite population. A sample s is drawn from the population, employing a design $p(s)$.

For a person labelled i, let.

$y_i = 1$ if i bears a sensitive attribute A say, drug addiction $= 0$, Otherwise.

$x_i = 1$ if i bears an innocuous attribute X, say birth in "August to December".

$= 0$ if born in "January to July".

The attributes A and X are supposed to be unrelated and independent. Here, $\frac{\sum_1^N x_i}{N}$, the proportion of persons bearing innocuous attribute X is known.

For i^{th} person we may define

$c_{00i} = [y_i = 0, y_i = 0]$, non-stigmatizing

$c_{01i} = [y_i = 0, x_i = 1]$, non-stigmatizing
$c_{10i} = [y_i = 1, x_i = 0]$, stigmatizing
$c_{11i} = [y_i = 1, y_i = 1]$, non-stigmatizing.
$d_i = c_{01i} \cup c_{10i} \cup c_{11i}$ is non-stigmatizing. Clearly, d_i is the union of the mutually exclusive events.

The "triangular model" consists in gathering a response as "1 or 0" from a sampled person i on either the direct event c_{00i} applicable to him/her or on d_i.

Our interest is to estimate θ.

Clearly, $\theta = \frac{1}{N}\left[\sum_1^N c_{10i} + \sum_1^N c_{11i}\right] = \frac{1}{N}\sum_1^N y_i$, the proportion of the population bearing the sensitive character A. Here,

$$\frac{1}{N}\sum_1^N c_{00i} = \frac{1-p}{N}\sum_1^N [y_i = 0] = (1-p)\text{Prob}(y = 0).$$

So, using Horvitz-Thompson (1952) method of estimation,
$\text{Prob}(y = 0) = \frac{1}{N}\sum_1^N [y_i = 0]$ may be unbiasedly estimated by $\frac{1}{N(1-p)}\sum_{i\in s}\frac{c_{00i}}{\pi_i}$.

Now, $\theta = \frac{1}{N}\left[\sum_1^N d_i - \sum_1^N c_{01i}\right]$.

Here, $\frac{1}{N}\sum c_{01i} = \frac{1}{N}\sum[y_i = 0, x_i = 1] = \frac{p}{N}\sum[y_i = 0] = p\text{Prob}(y = 0)$.

So, an unbiased estimator for θ is

$$\hat{\theta} = \frac{1}{N}\sum_{i\in s}\frac{d_i}{\pi_i} - \frac{p}{(1-p)}\frac{1}{N}\sum_{i\in s}\frac{c_{00i}}{\pi_i} = \frac{1}{N}\sum_{i\in s}\frac{e_i}{\pi_i},$$

where $e_i = d_i - \frac{p}{(1-p)}c_{00i}$.

For variance and the estimate of its variance, Horvitz-Thompson form of variance or Yates-Grundy form of variance and the related estimates may be calculated.

Modifications for Crosswise Model

Chaudhuri (2012) illustrated the modifications for crosswise non-randomized model where the sample is drawn by general sampling scheme. A "non-randomized response" from a sampled i is taken either, through a diagonal or the reverse diagonal and hence crosswise, given as

$$b_i = [y_i = 0, x_i = 0] \cup [y_i = 1, x_i = 1]$$

or

$$g_i = [y_i = 0, x_i = 1] \cup [y_i = 1, x_i = 0].$$

So,

$$\frac{1}{N}\sum_1^N b_i = (1-p)P(y=0) + pP(y=1)$$

$$\frac{1}{N}\sum_1^N g_i = pP(y=0) + (1-p)P(y=1).$$

Combining these, we have

$$(2p-1)\mathrm{Pr}(y=1) = \frac{1}{N}\left[p\sum b_i - (1-p)\sum g_i\right].$$

So, an unbiased estimator for $\theta = P(y=1)$ is

$$\hat{\theta} = \frac{1}{N}\frac{1}{(2p-1)}\left[\sum_{i\in s}(pb_i - (1-p)g_i)\right)] = \frac{1}{N}\sum_{i\in s}\frac{f_i}{\pi_i}.$$

Here,

$$f_i = \frac{[pb_i - (1-p)g_i]}{2p-1}$$

with $p \neq \frac{1}{2}$.

Variance and unbiased variance estimator formulas may be calculated as mentioned in triangular model.

Takahasi and Sakasegawa (1977) narrated a novel procedure, called implicit randomized response (IRR) technique avoiding any particular RR device unlike most researchers in this field.

Implicit Randomized Response (IRR) Device

Let $U = (1, \ldots, i, \ldots, N)$ denote a population of N persons,

$$y_i = \begin{cases} 1 & \text{if } i\text{th person bears a sensitive feature } A \\ 0 & \text{if } i\text{th person bears } A^C, \text{ the complement of } A, \quad i \in U. \end{cases}$$

Our problem is to estimate $Y = \sum_{i=1}^N y_i$ or equivalently $\theta_A = \frac{Y}{N} = \frac{\sum_{i=1}^N y_i}{N}$. Of course the values of $y_i's, i \in U$ are unknown. Pal (2007a) extended Takahasi and Sakasegawa's implicit randomized response (IRR) device for the samples drawn by any general unequal probability-sampling scheme.

Let

$$y_{iV} = \begin{cases} 1 & \text{if } i\text{th person bears } A \text{ and likes Violet most among three colours Violet, Blue, Green} \\ 0 & \text{else} \end{cases}.$$

Similarly, y_{iB}, y_{iG} are defined for the colours blue and green and $y_i = y_{iV} + y_{iB} + y_{iG}$, assuming one cannot like at a time more than one of these three colours.

Let us write $\theta_A = \theta_{AV} + \theta_{AB} + \theta_{AG}$., where $\theta_{AV} = \frac{1}{N}\sum_1^N y_{iV}$; $\theta_{AB} = \frac{1}{N}\sum_1^N y_{iB}$ and $\theta_{AG} = \frac{1}{N}\sum_1^N y_{iG}$.

Similarly, $\theta_{A^c} = (1 - \theta_A) = \theta_{A^cV} + \theta_{A^cB} + \theta_{A^cG}$. Three independent samples s_1, s_2 and s_3 are drawn from U with a suitable sampling design P each.

IRR device for the sample s_1

Let the people in a selected sample s_1 be requested to give independently two responses following instructions presented as in the sheet ST1 below.

ST1

(I)	A	A^C
V	0	1
B	1	0
G	1	0

(II)	A	A^C
V	1	0
B	0	1
G	1	0

(III)	A	A^C
V	1	0
B	1	0
G	0	1

He/she is to choose with probability p_1, p_2 ($0 < p_j < 1$, $j = 1, 2$) and $p_3 = 1 - p_1 - p_2$ ($0 < p_3 < 1$) indicated below one of these three sheets I-III of ST1 respectively and correctly note and report the indicated number 1/0 according to bearing one of A, A^c and preferring one of the three colours V, B, G, as applicable to himself/herself.

Denoting the two independent responses of ith individual of s_1 by r'_{1i} and r''_{1i}, we may define $r_{1i} = \frac{r'_{1i} + r''_{1i}}{2}$.

By E_R and V_R, we shall denote the expectation and variance operators with respect to this IRR device as above.

Thus,

$$E_R(r_{1i}) = E_R\left(r'_{1i}\right) = E_R\left(r''_{1i}\right)$$
$$= p_1(1 - y_{iV} + y_{iB} + y_{iG}) + p_2(y_{iV} + 1 - y_{iB} + y_{iG})$$
$$+ p_3(y_{iV} + y_{iB} + 1 - y_{iG}),$$
$$= M_{1i}, say, i \in s_1.$$

As in Chaudhuri (2001), two independent responses are required for unbiased estimation of the variance.

Let $v_R(r_{1i}) = \frac{(r'_{1i} - r''_{1i})^2}{4}$, $i \in s_1$.

Then, $E_R v_R(r_{1i}) = V_R(r_{1i})(say)$.

IRR Device for the Sample s_2

Following the same sampling design P as before, let a second sample s_2 be independently drawn from U again. A person in s_2 is requested to choose (I/), (II/) or (III/) below with probabilities p_1, p_2 or p_3 where $0 < p_j < 1, j = 1, 2$ and $p_3 = 1 - p_1 - p_2 (0 < p_3 < 1)$.

Let a sampled person i of s_2 be asked to report independently r'_{2i} and r''_{2i} defined similarly as r'_{1i} and r''_{1i} following similar instructions as in s_1 but noting the following changed structure, namely ST2.

$$ST2$$

(I′)

	A	A^C
V	1	0
B	1	0
G	0	1

(II′)

	A	A^C
V	0	1
B	1	0
G	1	0

(III′)

	A	A^C
V	1	0
B	0	1
G	1	0

Let $r_{2i} = \frac{r'_{2i} + r''_{2i}}{2}, v_R(r_{2i}) = \frac{(r'_{2i} - r''_{2i})^2}{4}, i \in s_2$.
Then,

$$E_R(r_{2i}) = p_1(y_{iV} + y_{iB} + 1 - y_{iG}) + p_2(1 - y_{iV} + y_{iB} + y_{iG})$$
$$+ p_3(y_{iV} + 1 - y_{iB} + y_{iG})$$
$$= E_R(r'_{2i}) = E_R(r''^{2i}) = M_{2i}(say),$$

$$E_R v_R(r_{2i}) = V_R(r_{2i}), , i \in s_2.$$

For the sample s_3, let a third independently drawn sample from U according to the same design P be s_3.

Here also the respondents select one of the structures (I//), (II//) or (III//) with probabilities p_1, p_2 or p_3 as described earlier.

Let every sampled respondent in it be requested to yield independent variables r'_{3i} and $r''_{3i}, i \in s_3$ similarly as r'_{2i}, r''_{2i}, etc., with only the following 'changed structure ST3'.

ST3

	(I$''$)	
	A	**A^C**
V	1	0
B	0	1
G	1	0

	(II$''$)	
	A	**A^C**
V	1	0
B	1	0
G	0	1

	(III$''$)	
	A	**A^C**
V	0	1
B	1	0
G	1	0

Then, as in the earlier two cases we generate $r_{3i} = \frac{r'_{3i}+r''_{3i}}{2}$ and $v_R(r_{3i}) = \frac{(r'_{3i}-r''_{3i})^2}{4}$, $i \in s_3$ such that

$$E_R(r_{3i}) = p_1(y_{iV} + 1 - y_{iB} + y_{iG}) + p_2(y_{iV} + y_{iB} + 1 - y_{iG})$$
$$+ p_3(1 - y_{iV} + y_{iB} + y_{iG}) = M_{3i}(say).$$

Then, $E_R v_R(r_{3i}) = V_R(r_{3i})$, $i \in s_3$.

Next to find unbiased estimators of proportion θ_A and their variance estimators, we proceed as given below.

For the sample s_1, proceeding with Horvitz-Thompson method of estimation and writing E_P, V_P as expectation, variance operators in respect of the design P, let us define

$$\hat{\theta}_1 = \frac{1}{N} \sum_{i \in s_1} \frac{r_{1i}}{\pi_i} \text{ such that } E_P(\hat{\theta}_1) = \frac{1}{N} \sum_{1}^{N} r_{1i}.$$

Also writing $E = E_p E_R$, $V = V_p E_R + E_p V_R$ as the overall expectation and variance operators, we have

$$E(\hat{\theta}_1) = p_1(\theta_{A^c V} + \theta_{AB} + \theta_{AG}) + p_2(\theta_{AV} + \theta_{A^c B} + \theta_{AG}) + p_3(\theta_{AV} + \theta_{AB} + \theta_{A^c G}),$$

and $V(\hat{\theta}_1) = E_P V_R(\hat{\theta}_1) + V_P E_R(\hat{\theta}_1)$.

Similarly, for samples s_2 and s_3, $\hat{\theta}_2$ and $\hat{\theta}_3$ are defined. Then,

$$E(\hat{\theta}_2) = p_1(\theta_{AV} + \theta_{AB} + \theta_{A^c G}) + p_2(\theta_{A^c V} + \theta_{AB} + \theta_{AG}) + p_3(\theta_{AV} + \theta_{A^c B} + \theta_{AG}),$$

and $E(\hat{\theta}_3) = p_1(\theta_{AV} + \theta_{A^c B} + \theta_{AG}) + p_2(\theta_{AV} + \theta_{AB} + \theta_{A^c G}) + p_3(\theta_{A^c V} + \theta_{AB} + \theta_{AG})$.

Let finally,

$$\hat{\theta}'_A = \hat{\theta}_1 + \hat{\theta}_2 + \hat{\theta}_3$$

be defined to yield

$$E(\hat{\theta}'_A) = E(\hat{\theta}_1) + E(\hat{\theta}_2) + E(\hat{\theta}_3)$$
$$= (1 + \theta_A)(p_1 + p_2 + p_3) = 1 + \theta_A.$$

Then, a proposed unbiased estimator for θ_A is.
$\hat{\theta}_A = \hat{\theta}'_A - 1 = \hat{\theta}_1 + \hat{\theta}_2 + \hat{\theta}_3 - 1$ and a proposed unbiased estimator for $V(\hat{\theta}_A)$ is
$v(\hat{\theta}_A) = \hat{V}(\hat{\theta}_1) + \hat{V}(\hat{\theta}_2) + \hat{V}(\hat{\theta}_3)$.

We shall now separately present our formulae for $\widehat{V}(\hat{\theta}_j)$, $j = 1, 2, 3$ following essentially the approach of Chaudhuri (2001). The related variance and its estimate are obtained by Horvitz-Thompson method of estimation.

11.3 Item Count Techniques (ICT)

Item count technique (ICT) was originally proposed by Miller (1984). It is also called the unmatched count technique. One advantage of the item count technique over the randomized response technique is that no randomized device is required. However, two groups of respondents are generated via randomization. In ICT, a questionnaire contains statements on various innocuous characteristics along with a statement on the stigmatizing characteristic. The sampled individuals are requested to answer the total number of statements which hold true for them in a questionnaire.

11.3.1 ICT for General Sampling Design

Let $U = (1, \ldots, i, \ldots, N)$ denote a population of a known number of N units.

Let $\theta_i = 1$ if the ith person bears a sensitive characteristic $A = 0$, he/she bears A^C, the complement of A.

Our problem is to estimate $\theta_A = \frac{\sum_{i=1}^{N} \theta_i}{N}$. In practice often the values of θ_i's, $i \in U$ are non-ascertainable.

To estimate the unknown proportion θ_A, we may alternatively employ item count technique. Following Chaudhuri and Christofides (2007), in the item count method two nearly identical lists of behaviours are developed of which G (>1) items or behaviours of both the lists are innocuous and exactly same. The $(G + 1)^{st}$ item of the first list stands for either the stigmatizing feature A or any fresh innocuous item (say F) or both of them (i.e. $A \cap F$).

In the other list, the $(G + 1)^{st}$ item stands for either the complement of the stigmatizing question (A^C) or the complement of the fresh item (say (F^C)) or both the complement of the stigmatizing and the complement of the fresh innocuous item (i.e. $(A^C \cap F^C)$).

$$\text{Let } F_i = \begin{cases} 1 & \text{if the } i\text{th person bears the fresh innocuous item } F \\ 0 & \text{if the } i\text{th person does not bear the item F} \end{cases},$$

and

$$\theta_i F_i = \begin{cases} 1 & \text{if the } i\text{th person bears both the character } A \text{and the fresh innocuous item } F \\ 0 & \text{if the } i\text{th person does not bears either of the character } A \text{ and the item } F, \text{ as above} \end{cases}.$$

Let two independent samples s_1 and s_2 be drawn from U with a common sampling design P. Respondents of the sample s_1 are asked to simply provide the number of behaviours, say $y_i (i = 1, 2, \ldots, N)$ which apply to them (without indicating the specific behaviours) among the $(G + 1)$ items of the first list. Every unit in s_2 is presented with the second list of $(G + 1)$ items. The number of behaviours applicable for an individual j of the second sample is x_j (say). Following Chaudhuri and Christofides (2007) and using the Horvitz-Thompson estimation (1952) method, the unknown proportion θ_A may be estimated by $\hat{\theta}_A(1) = \frac{1}{N} \sum_{i \in s_1} \frac{y_i}{\pi_i} - \frac{1}{N} \sum_{j \in s_2} \frac{x_j}{\pi_j} + 1 - \theta_F$ where $\theta_F = \frac{1}{N} \sum_{k=1}^{N} F_k$ (which is known) and π_i, π_j s are the first-order inclusion probabilities.

Writing the second-order inclusion probabilities for a pair of units i and i' $(i \neq i')$ as $\pi_{ii'}$ and using Horvitz-Thompson (1952) method of estimation, an unbiased estimator of $V(\hat{\theta}_A(1))$ is

$$v_P(\hat{\theta}_A(1)) = \sum_{i < i' \in s_1} \sum \left(\frac{\pi_i \pi_{i'} - \pi_{ii'}}{\pi_{ii'}} \right) \left(\frac{y_i}{\pi_i} - \frac{y_{i'}}{\pi_{i''}} \right)^2$$
$$+ \sum_{j < j' \in s_2} \sum \left(\frac{\pi_j \pi_{j'} - \pi_{jj'}}{\pi_{jj'}} \right) \left(\frac{x_i}{\pi_j} - \frac{x_j}{\pi_{j'}} \right)^2.$$

11.3.2   ICT for Quantitative Sensitive Characteristic

Let μ_y be the population mean of the quantitative stigmatizing characteristic of interest. For example, the stigmatizing characteristic could be the number of abortions induced. To estimate μ_y, two independent random samples s_1 and s_2 of sizes n_1 and n_2 are drawn with replacement from the population. Each one of the participants from the first sample is presented with a questionnaire (list) of $G+1$ items, with G of those related to non-stigmatizing items and one related to the stigmatizing sensitive characteristic A.

All the items are quantitative in nature. The respondent observes each one of the items and writes down (for his/her own convenience) the value applicable to him/

her (for each one of the items). He/she reports the total of all the values together. A participant from the second sample (s_2) is presented with the list of the G non-stigmatizing items, exactly the same as the ones included in the questionnaire for the first sample s_1. Again the participant is to report the total of the values applicable to him/her, without reporting the individual values.

Here, all the G + 1 items are independent of one another. Let $x_1, x_2, \ldots, x_G$ denote the variables representing the G non-sensitive items with $\mu_{x_1}, \ldots \mu_{x_2}, \ldots \mu_{x_G}$, and $\sigma_{x_1}^2, \sigma_{x_2}^2, \ldots, \sigma_{x_G}^2$ denoting the population means and variances, respectively.

Let $\overline{T}^{(1)}, \overline{T}^{(2)}$ denote the averages of the values reported by the respondents of sample 1 and sample 2, respectively. Then, μ_y is unbiasedly estimated by $\overline{T}^{(1)} - \overline{T}^{(2)}$ with variance

$$\frac{\sigma_y^2}{n_1} + \left(\left(\sum_{i=1}^{G} \sigma_{x_i}^2 \right) \left(\frac{1}{n_1} + \frac{1}{n_2} \right) \right).$$

Chaudhuri and Christofides' (2013) version of ICT for estimating finite population mean of a sensitive quantitative variable also mandates selection of two independent samples. Removing the restrictions, Shaw (2015, 2016) modified both the ICTs.

11.3.3 Option Between RR and ICT in Surveys

To take care of the respondents opting for RR and those preferring ICT to RR, Pal (2007a, 2007b) developed an optional method providing each sampled individual i in s the option to either answer an RR or an ICT questionnaire, without revealing the choice to the investigator. Each respondent i, in the population, bears an unknown probability to opt for an RR and with the complementary probability for an ICT.

To estimate the proportion of people bearing a sensitive character several RR techniques are there in the literature. But here to cover the above item count values, a special RR technique has been introduced. Let u be a random variable following a discrete uniform distribution (0, G), i.e. Prob $[u = k] = \frac{1}{(G+1)}, k = 0, 1, \ldots, G$. Then, the random variable u takes the values $0,1,2,\ldots,G$. As his/her RR response, the ith individual of the sample s_1 will report $I_i = \theta_i + u_i$ where u_i is the value of the random variable u. By E_R and V_R, we shall denote the expectation and variance operator with respect to the RR device. Thus, $E_R(I_i) = \theta_i + \alpha$, where α is the mean of the random variable u. Clearly, $\alpha = G/2$.

We suppose that θ_i is not ascertainable for a person i in a sample. Here, an RR may be procured as $r_i = I_i - \alpha$ such that.

(i) $E_R(r_i) = \theta_i$,
(ii) $V_R(r_i) = V_i(RR)(> 0)$,
(iii) r_i s are independent over i in U and

(iv) there exists v_i ascertainable from RR's such that $E_R(v_i(RR)) = V_i(RR) = \frac{G(G+2)}{12}, i \in U$.

Writing E_P, V_P as expectation and variance operators in respect of the design P, let us write $E = E_P E_R = E_R E_P$ and $V = V_P E_R + E_P V_R = E_R V_P + V_R E_P$, where E and V are the overall expectation and the variance operators, respectively.

Here, our proposed estimator is

$$\hat{\theta}_A(2) = \frac{1}{N} \sum_{i \in s_1} \frac{r_i}{\pi_i}.$$

It may be estimated by

$$v(\hat{\theta}_A(2)) = \sum_{i<j \in s_1} \sum \left(\frac{\pi_i \pi_j - \pi_{ij}}{\pi_{ij}} \right) \left(\frac{r_i}{\pi_i} - \frac{r_j}{\pi_j} \right)^2 + \sum_{i \in s_1} \frac{v_i(RR)}{\pi_i}.$$

Here, each respondent selected in the sample has either of the following two options:

(i) he/she can provide the randomized response (RR)

or (ii) the respondent can report the number of items applicable to him/her using the item count technique.

Each individual will give out his/her item count value without disclosing whether he/she is doing so or follow the instructions as described in the RR device. Two independent samples s_1 and s_2 have been drawn for our proposed Optional technique. It is supposed that ith respondent of the sample s_1 chooses to give out the randomized response (RR) I_i with probability $c_i (0 \le c_i \le 1)$ (which is unknown) or the item count value (y_i) with probability $(1-c_i), i \in U$. For his/her RR value, ith respondent will report the value $I_i = \theta_i + u_i$, where u_i is the value of the random variable u. Two independent responses may be gathered from each individual.

Let the first response of the ith person of the sample s_1 denoted as z_{1i} be defined as.

$z_{1i} = \theta_i + u_i = I_i$ with an unknown positive probability c_i or $= y_i$, with the complementary probability $(1- c_i), i = 1, 2, ..., N$.

A second independent response from each respondent i of the sample s_1 is also requested either to apply the RR device or to apply the item count technique. Denoting the second independent response of a sampled person i as z_{2i}, we may write $z_{2i} = \theta_i + u'_i = I'_i$ with an unknown positive probability c_i or $= y_i$, with the complementary probability $(1- c_i), i = 1, 2, \ldots, N$, where u'_i is the value of the random variables u with a fresh independent draw.

Writing E_R and V_R as the operators for expectation and variance with respect to any RR device, we may write $E_R(z_{1i}) = (\theta_i + \alpha)c_i + (1 - c_i)y_i = E_R(z_{2i})$.

Let $z_i = \frac{z_{1i}+z_{2i}}{2}$ then $E_R(z_i) = (\theta_i + \alpha)c_i + (1 - c_i)y_i = E_R(z_{1i}) = E_R(z_{2i})$.

Defining $v_i = \frac{(z_{1i}-z_{2i})^2}{4}$, we also may write $E_R(v_i) = V_R(z_i) = V_i = V_R(z_{1i}) = V_R(z_{2i})$.

Each respondent selected in the sample s_2 has either of the following two options:

(i) to make our procedure convenient the respondent numbered i of the sample s_2 is being requested to report the value $1 + u_i - F_i$ as he/she chooses RR method or (ii) he/she will report the value x_i (of Sect. 11.2) as obtained by item count method. A second independent response is also requested from ith respondent of the sample s_2. Denoting their two responses by z'_{1i} and z'_{2i}, we may write

$$E_R(z'_{1i}) = (1 + \alpha - F_i)c_i + (1 - c_i)x_i = E_R(z'_{2i}).$$

Let $z'_i = \frac{(z'_{1i} + z'_{2i})}{2}$, $v'_i = \frac{(z'_{1i} - z'_{2i})^2}{4}$ then $E_R(z'_i) = E_R(z'_{1i}) = E_R(z'_{2i})$, and $E_R(v'_i) = V_R(z'_i) = V'_i = V_R(z'_{1i}) = V_R(z'_{2i})$.

In a practical situation G is only 4 or 5. So for both the responses z_{1i} and z_{2i} (or, z'_{1i} and z'_{2i}) of a particular individual the RR values may be same.

Define $e_1 = \frac{1}{N} \sum_{i \in s} \frac{z_i}{\pi_i}$ and $e_2 = \frac{1}{N} \sum_{i \in s} \frac{z'_i}{\pi_i}$. Now,

$$E(e_1) = E_P E_R(e_1) = \frac{1}{N} \sum_{i=1}^{N} [(\theta_i + \alpha)c_i + (1 - c_i)y_i]$$

$$E(e_2) = E_P E_R(e_2) = \frac{1}{N} \sum_{i=1}^{N} [(1 + \alpha - F_i)c_i + (1 - c_i)x_i].$$

Writing $A^C \cup F^C = (A \cap F)^C$ and using De-Morgan's law for the indicator function, we may write

$$E(e_1) - E(e_2)$$

$$= \frac{1}{N} [(\theta_i + \alpha)c_i - (1 + \alpha - F_i)c_i + (1 - c_i)\{\theta_i + F_i - \theta_i F_i - (1 - \theta_i F_i)\}]$$

$$= \frac{1}{N} \left[\sum \theta_i + \sum F_i - N \right] = \theta_A + \theta_F - 1,$$

where $\theta_F = \frac{\sum_{i=1}^{N} F_i}{N}$ which is known. Hence, our proposed estimator is $\hat{\theta}_A(3) = e_1 - e_2 - \theta_F + 1$. The variance of the proposed estimator is

$$V(\hat{\theta}_A(3)) = V(e_1) + V(e_2), \text{ where}$$
$$V(e_1) = E_R V_P(e_1) + V_R E_P(e_1)$$
$$= E_R V_P(e_1) + \sum V_i,$$

and

$$V(e_2) = E_R V_P(e_2) + \sum V'_i.$$

Our proposed unbiased estimator of $V(\hat{\theta}_A(3))$ is

$$v(\hat{\theta}_A(3)) = v_P(e_1) + \sum_{i \in s_1} \frac{v_i}{\pi_i} + v_P(e_2) + \sum_{i \in s_1} \frac{v_i'}{\pi_i}.$$

Then, $v(e_1)$ and $v(e_2)$ may be written as

$$v_P(e_1) = \sum_{i < j \in s_1} \sum \left(\frac{\pi_i \pi_j - \pi_{ij}}{\pi_{ij}}\right) \left(\frac{z_i}{\pi_i} - \frac{z_j}{\pi_j}\right)^2,$$

and

$$v_P(e_2) = \sum_{i < j \in s_2} \sum \left(\frac{\pi_i \pi_j - \pi_{ij}}{\pi_{ij}}\right) \left(\frac{z_i'}{\pi_i} - \frac{z_j'}{\pi_j}\right)^2.$$

Shaw and Pal (2021) narrated estimation methods of optional randomized response device allowing the sampled individuals to either respond directly or provide an RR or answer to an ICT questionnaire, according to their choices.

Imai (2011) proposed nonlinear least squares and maximum likelihood estimators for efficient multivariate regression analysis with the item count technique.

11.4   Nominative Technique

Another alternative to randomized response is the "nominative technique." In this method, each of the sampled persons is asked to indicate, i.e. nominate as many of his/her acquaintances having the stigmatizing characteristic. For example, he/she may be asked to say how many people he/she is aware of making illegal drug use. The primary purpose of the nominative technique is to minimize respondent denial of socially undesirable behaviour. Another possible advantage is achieving coverage of "hard-to-reach" deviant population groups.

11.4.1   Miller's (1985) Nominative Technique

In order to estimate the proportion θ of people in a community bearing a sensitive attribute A, on gathering relevant sensitive data by indirect questioning, Miller (1985) introduced a new method called the "nominative method." It was developed for the purpose of estimating heroin prevalence in the general population. It involves asking respondents to report on their close friend's heroin use. In it, from a relevant community constituting a population, an SRSWR is chosen and every sampled person is requested to identify the number of his/her acquainted persons known to bear the stigmatizing attribute A. The nominative technique is based upon the following proposition: If each member of the population reports the number (0, 1,

2, 3...) of close friends who have used heroin, then, with appropriate correction for duplication, it is possible to derive an accurate count of the number of heroin users in the population.

The two key items in the nominative question series are (in essence):

(i) So far as you know, how many of your close friends have ever used heroin? Just count the ones that you know for sure has used it. Then, for each of the interviewee's heroin-using close friends.

(ii) How many of this person's other close friends (besides yourself) know that he (or she) has used heroin?

The information gathered in Question B allows an appropriate "weight" (or correction for duplication) to be attached to each report of a heroin-using friend. This weight is the inverse of the total number of persons in the population who are eligible to report that particular heroin user. Specifically, the fractional weight that must be attached to the jth interviewee's report of the ith heroin user is: $\frac{1}{1+B_{ij}}$. This weight corrects for the fact some heroin users will be reported by two, three, four, or more close friends. But duplicated count should be focused. The grand total of weighted counts of users would be the same as the total number of users in the population.

Thus, in the practical version of the nominative technique, each interviewee or respondent has either a "0" score for no users known or a "1" score for a report of a particular heroin user known. This particular user is either the only user that the respondent knows or is the "random one" selected from all users known to that respondent. This particular user, who has a number of characteristics such as sex, age and regency of use, bears two data weights:

A_j refers "weights up" the jth respondent's report of this heroin user to stand for the total number of users that he or she has mentioned.

$\frac{1}{1+B_{ij}}$ is the correction for duplicate reports of the same user.

The combined weight factor for each score of "1" (i.e. for each respondent who knows a heroin user) is $A_j\left(\frac{1}{1+B_j}\right)$.

(The i subscript has been dropped).

Using this combined weight factor—and summing reports across all respondents in a complete census of a given population—yields an estimate of the total number of heroin users in that population.

11.4.2 Chaudhuri and Christofides's (2008) Modification on Nominative Technique

The nominative methodology can be thought of as an application of network sampling introduced by Thompson (1992) and further developed by Thompson and Seber (1996) and Chaudhuri (2000). By exploiting the Network Sampling technique and its analogous link with Miller's ideas, Chaudhuri and Christofides (2008) showed how Miller's ideas may be exploited to estimate θ on taking a sample by a suitably general scheme.

To develop the mathematical foundation of the method, in a population of size N, let r_{ij} for $i \neq j$ take the value 1 if the jth participant reports that the ith member of the population has the stigmatizing attribute and 0 otherwise. Clearly, $\sum_1^N r_{ij}$ represents the number of people reported by the jth participant as belonging to the stigmatizing group, and similarly, $\sum_1^N r_{ij}$ represents the number of times that the ith member of the population is reported as having the sensitive characteristic.

Finally, $\sum_{i=1}^N \sum_{j=1, j \neq i}^N r_{ij}$ gives the number of reports that persons belong to the stigmatizing group. Then, the total number of people in the community having the stigmatizing attribute is given by

$$T = \sum_{i=1}^N \sum_{j=1, j \neq i}^N \left(r_{ij} / \sum_{k=1}^N r_{ik} \right),$$

where the quantity $r_{ij} / \sum_{k=1}^N r_{ik}$ is taken to be zero if $\sum_{k=1}^N r_{ik}$ is equal to zero.

Let A_j denote the number of people belonging to the stigmatizing group reported by the jth participant and let B_j denote the number of close friends of nominees reported by j who know that this individual belongs to the stigmatizing group. Let $x_j = \frac{A_j}{1+B_j}$.

Here $\hat{\theta} = \frac{1}{n} \sum_1^n x_{j\cdot}$.

After nearly six decades since the technique was invented, many improvements of the randomized response techniques and other indirect questioning techniques have appeared in the literature. Perri et al. (2021) proposed a procedure to detect the presence of liars in sensitive surveys which allows researchers to evaluate the impact of untruthful responses on the estimation of the prevalence of a sensitive attribute, making use of direct and indirect questions. They applied their proposal to the Warner randomized response method, the unrelated question model, the item count technique, the crosswise model and the triangular model.

Recently, Truong-Nhat Le (2023) presented a review of RRTs from the work of Warner (1965) to the present. But they skipped Chaudhuri's (2001) modifications of RR and NRR theories in estimation for general sampling designs.

References

Chaudhuri, A. (2000). Network and adaptive sampling with unequal probabilities. *Calcutta Statistical Association Bulletin, 50*, 237–253.

Chaudhuri, A. (2001). Using randomized response from a complex survey to estimate a sensitive proportion in a dichotomous finite population. *Journal of Statistical Planning and Inference, 94*, 37–42.

Chaudhuri, A. (2012). Unbiased estimation of a sensitive proportion in general sampling by three nonrandomized response schemes. *Journal of Statistical Theory and Practice, 6*, 376–381.

Chaudhuri, A., & Christofides, T. C. (2007). Item count technique in estimating the proportion of people with a sensitive feature. *Journal of Statistical Planning and Inference, 137*, 589–593.

Chaudhuri, A., & Christofides, T. C. (2008). Indirecting questioning: How to rival randomized response techniques. *International Journal of Pure and Applied Mathematics, 43*, 283–294.

Chaudhuri, A., & Christofides, T. C. (2013). *Indirect questioning in sample surveys.* Springer-Verlag.

Christofides, T. C. (2009). Randomized response without a randomization device. *Advances and Applications in Statistics, 11*, 15–28.

Droitcour, J. A., Larson, E. M., & Scheuren, F. J. (2001). The three card method: estimating sensitive items with permanent anonymity of response. In *Proceedings of the Annual Meeting of the American Statistical Association.*

Horvitz, D. G., & Thompson, D. J. (1952). A generalization of sampling without replacement from finite universe. *Journal of the American Statistical Association, 47*, 663–685.

Imai, K. (2011). Multivariate regression analysis for the item count technique. *Journal of American Statistical Association, 106*(494), 407–416.

Miller, J. D. (1984). *A new survey technique for studying deviant behavior.* Ph.D. Thesis, The George Washington University

Miller, J. D. (1985). The nominative technique: A new method of estimating heroin prevalence. *NIDA Research Monograph, 54*, 104–124.

Pal, S. (2007a). Estimating the proportion of people bearing a sensitive issue with an option to item count lists and randomized response. *Statistics in Transition, 8*, 301–310.

Pal, S. (2007b). Extending Takahasi–Sakasegawa's indirect response techniques to cover sensitive surveys in unequal probability sampling. *Calcutta Statistical Association Bulletin, 59*(3–4), 265–276.

Perri, P., Manoli, E., & Christofides, T. C. (2022). Assessing the effectiveness of indirect questioning techniques by detecting liars. *Statistical Papers, 43*, 283–294.

Shaw, P. (2015). Estimating a finite population mean of a sensitive quantitative variable from a single probability sample by the item count technique. *Model Assisted Statistics and Applications, 10*, 411–419.

Shaw, P. (2016). Estimating a finite population proportion bearing a sensitive attribute from a single probability sample by item count technique. *Handbook of Statistics, 34*, 387–403. North-Holland, Elsevier B.V.

Shaw, P., & Pal, S. (2021). Estimating sensitive population proportion permitting options for various respondents' choices. *Statistics and Applications, 19*(2), 161–179.

Takahasi, K., & Sakasegawa, H. (1977). An RR technique without use of any randomizing device. *Annals of the Institute of Statistical Mathematics, 29*, 1–8.

Tan, M. T., Tian, G. L., & Tang, M. L. (2009). Sample surveys with sensitive questions: A nonrandomized response approach. *American Statistician, 63*, 9–16.

Thompson, S. K. (1992). *Sampling.* John Wiley & Sons.

Thompson, S. K., & Seber, G. A. (1996). *Adaptive sampling.* John Wiley & Sons.

Tian, G., & Tang, M. (2014). *Incomplete categorical data design: non-randomized response techniques for sensitive questions in surveys.* CRC Press.

Tian, G. L., Yu, J. W., Tang, M. L., et al. (2007). A new non-randomized model for analysis sensitive questions with binary outcomes. *Statistics in Medicine., 26*(23), 4238–4252.

Le, T.-N., Lee, S.-M., Tran, P.-L., & Li, C.-S. (2023). Randomized response techniques: a systematic review from the pioneering work of warner (1965) to the present. *Mathematics, 11*, 1718.

Yu, J. W., Tian, G. L., & Tang, M. L. (2008). Two new models for survey sampling with sensitive characteristic: Design and analysis. *Metrika, 67*, 251–263.

Wu, Q., & Tang, M. (2016). Non-randomized response model for sensitive survey with noncompliance. *Statistical Methods in Medical Research, 25*(6), 2827–2839.

Warner, S. L. (1965). Randomized response: A survey technique for eliminating evasive 1566 answer bias. *Journal of American Statistical Association, 60*, 63–69.

Chapter 12
An Epilogue

After Warner (1965) gave us his pioneering technique of randomized response technique (RRT), the subject has ramified into various directions. The Handbook of Statistics 34 (2016) tells about many of its aspects. Adhikari et al. (1984) discussed mainly about extension of general theory of survey sampling with direct response survey data to cover situations when the data are based on randomized response techniques gathered in diverse ways and some of the contemporary and immediately following literature also covered them.

This literature permitted data collection by general methods of sample selection, mainly with varying probabilities without replacement. But the main flow of the growth of RRT continued with the archaic SRSWR as the sole carrier of the accumulating materials with RR survey data. Since Warner himself started with SRSWR, most of his followers treated nothing but SRSWR alone. So, a large community of experts in RRT grew up exclusively dealing with SRSWR alone. Consequently budding researchers on RRT continued to be standard bearers of SRSWR in the context of RRs because the peer reviewers of their forthcoming research outputs were experts in SRSWR. Consequently, the theory could not progress in desirable ways to meet the requirements of experts in general survey sampling theory and practice. Eriksson (1973) pioneered treatment of RR data gathered by varying probability sampling but he had only a few takers. Chaudhuri (2011) through his monograph tried to spread a stimulus to all researchers in RRT to abandon SRSWR and take recourse to varying probability sampling. But till date, he has only a modicum of takers. Even today, many peer-reviewed papers in the area of RRT are predominantly based on SRSWR although his 2011 book is zealously referred to in the bibliographies.

In the present monograph in the Chap. 3, we demonstrate how to judge in practice if a complex design-based RR data may be more efficacious than the RR survey data procured through SRSWR or SRSWOR. In Chap. 4, we also comprehensively discuss how to decide on the sizes of samples to hit upon with equal or varying probabilities with or without replacement. Some RRTs are based on stratified samples, and in such, a context it is also considered relevant to decide how the strata sample-sizes are to be allocated. In classical survey sampling literature, everybody learns Neyman's sample-size allocation theory in stratified sampling without bothering about how the total sample-size is decided upon before contemplating its allocation to the various strata. Chaudhuri and Chakrabarti (2021) have clearly and elegantly discussed how to apply Chebyshev's inequality to hit upon the sample-size in SRSWR and SRSWOR and hence similarly work out sample-sizes in choosing the SRSWR or SRSWORs from various strata without bothering about allocation problem starting with a total sample-size at hand.

The authors of the present monograph fondly hope research outputs in plenty to follow from those who will be the avid readers of it. Hope it may not remain a mere wishful thought in the three of us. We sincerely believe our presentation has remained reader-friendly throughout the monograph's length and breadth.

References

Adhikari, A. K., Chaudhuri, A., & Vijayan, K. (1984). Optimum sampling strategies for randomized response trials. *International Statistical Review, 52*(2), 115–120.

Chaudhuri, A. (2011). *Randomized response and indirect questioning techniques in surveys.* CRC Press.

Chaudhuri, A., & Chakraborty, C. (2021). Rationalizing sample-size allocation in stratified sampling. *Journal of the Indian Society of Agricultural Statistics, 75*(3), 215–220.

Chaudhuri, A., Christofides, T. C., & Rao, C. R. (2016). *Handbook of statistics, data gathering, analysis and protection of privacy through randomized response techniques: Qualitative and quantitative human traits* (Vol. 34). Elsevier.

Eriksson, S. A. (1973). A new model for randomized response. *International Statistical Review, 41,* 101–113.

Warner, S. L. (1965). Randomized response: A survey technique for eliminating evasive answer bias. *Journal of American Statistical Association, 60,* 63–69.

Printed in the United States
by Baker & Taylor Publisher Services